SpringerBriefs in Computer Science

SpringerBriefs present concise summaries of cutting-edge research and practical applications across a wide spectrum of fields. Featuring compact volumes of 50 to 125 pages, the series covers a range of content from professional to academic. Typical topics might include:

- A timely report of state-of-the art analytical techniques
- A bridge between new research results, as published in journal articles, and a contextual literature review
- A snapshot of a hot or emerging topic
- An in-depth case study or clinical example
- A presentation of core concepts that students must understand in order to make independent contributions

Briefs allow authors to present their ideas and readers to absorb them with minimal time investment. Briefs will be published as part of Springer's eBook collection, with millions of users worldwide. In addition, Briefs will be available for individual print and electronic purchase. Briefs are characterized by fast, global electronic dissemination, standard publishing contracts, easy-to-use manuscript preparation and formatting guidelines, and expedited production schedules. We aim for publication 8–12 weeks after acceptance. Both solicited and unsolicited manuscripts are considered for publication in this series.

**Indexing: This series is indexed in Scopus, Ei-Compendex, and zbMATH **

Lin Zhang • Ming Xiao • Zicun Wang •
Wanbin Tang

AI-enabled Spectrum Sharing

Recent Advances In Wireless Edge Networks

 Springer

Lin Zhang
National Key Laboratory of Wireless
Communications
University of Electronic Science and
Technology of China
Chengdu, Sichuan, China

Zicun Wang
National Key Laboratory of Wireless
Communications
University of Electronic Science and
Technology of China
Chengdu, Sichuan, China

Ming Xiao
Division of Information Science and
Engineering
KTH Royal Institute of Technology
Stockholm, Sweden

Wanbin Tang
National Key Laboratory of Wireless
Communications
University of Electronic Science and
Technology of China
Chengdu, Sichuan, China

ISSN 2191-5768 ISSN 2191-5776 (electronic)
SpringerBriefs in Computer Science
ISBN 978-981-97-7643-6 ISBN 978-981-97-7644-3 (eBook)
https://doi.org/10.1007/978-981-97-7644-3

This Springer imprint is published by the registered company Springer Nature Singapore Pte Ltd.
The registered company address is: 152 Beach Road, #21-01/04 Gateway East, Singapore 189721,
Singapore

If disposing of this product, please recycle the paper.

Preface

Recent years have witnessed the rapid developments of wireless communication ecosystems including fundamental theory breakthroughs, manufacture capability improvements, as well as the explosively increasing wireless end devices and service demands. Wireless edge networks aim to provide first/last-mile wireless connections between access points and diversified wireless end devices, and is predicted to be extremely large-scale and heterogeneous in the next generation. By starting with a brief overview of existing spectrum sharing techniques, this Springer Brief focuses on the artificial intelligence (AI)-enabled spectrum sharing advances for three typical scenarios in future wireless edge networks, including opportunistic spectrum sharing, centralized spectrum sharing, and distributed spectrum sharing. For each scenario, novel AI-enabled transmissions are designed and the corresponding performance is extensively evaluated. The results show that the designed AI-enabled spectrum sharing methods can enhance the spectrum efficiency in wireless edge networks.

Chengdu, China   Lin Zhang
Stockholm, Sweden   Ming Xiao
Chengdu, China   Zicun Wang
Chengdu, China   Wanbin Tang

Contents

Acronyms

AI	Artificial Intelligence
AP	Access Point
BS	Base Station
CNN	Convolutional Neural Network
CR	Cognitive Radio
CSI	Channel State Information
DDPG	Deep Deterministic Policy Gradient
DL	Deep Learning
DNN	Deep Neural Network
DPG	Deterministic Policy Gradient
DQN	Deep Q-Network
DRL	Deep Reinforcement Learning
FP	Fractional Programming
HetNet	Heterogeneous Network
IID	Independent and Identically Distributed
IoT	Internet of Things
ITU	International Telecommunication Union
MAIAC	Multi-Agent Independent Actor-Critic
MASC	Multiple-Actor-Shared-Critic
MCS	Modulation and Coding Scheme
MDP	Markov Decision Process
MSE	Mean Square Error
PER	Packet Error Rate
PG	Policy Gradient
PT	Primary Transmitter
RL	Reinforcement Learning
SE	Spectrum Efficiency
SER	Symbol Error Rate
SINR	Signal to Interference and Noise Ratio
SNR	Signal-to-Noise Ratio
ST	Secondary Transmitter

UCB	Upper Confidence Bandit
UE	User Equipment
WMMSE	Weighted Minimum Mean Square Error
ZB	Zettabytes

Chapter 1
Introductions and Preliminaries

Abstract This chapter introduces the importance of spectrum sharing in wireless edge networks. With the rapid development of wireless communications, global mobile data traffic has explosively grown, resulting in the congestion in spectrum resources. Therefore, spectrum sharing has emerged as a key technique to address this issue. In this chapter, we introduce *cognitive radio* (CR) and *artificial intelligence* (AI) as two representative techniques for spectrum sharing. Particularly, we cover general reinforcement learning (RL), deep Q-network (DQN) and *deep deterministic policy gradient* (DDPG) as essential preliminaries for AI-enabled spectrum sharing. Eventually, the structure of the Brief is outlined.

Keywords Spectrum sharing · Cognitive radio · Deep learning · Reinforcement learning

1.1 Introductions

With the rapid developments of wireless communication theories and technologies, the global mobile data traffic explosively grows in recent years. It is predicted by the *International Telecommunication Union* (ITU) that the growth of the overall global mobile data traffic will reach astonishingly five *zettabytes* (ZB) per month by 2030 [1, 2]. Wireless edge network is deployed to provide the first/last-mile connections between access points and diversified end terminals, such as smart phones and *Internet of Things* (IoT) devices, and carries the majority of the mobile data traffic.

Spectrum resource plays an important role for mobile communications in wireless edge networks. With the demands of varieties of wireless services grow, the spectrum resource becomes extremely congested. Conventionally, the spectrum resource is officially managed by the regulators which aim to utilize efficiently the scarce spectrum with fairness among different stakeholders to some extent [1, 2]. To accommodate the ever-increasing wireless services, it is demanding and inefficient for the regulators to reserve dedicated spectrum for different wireless services. Alternatively, the spectrum sharing is widely suggested to achieve the coexistence and symbiosis among different wireless services on the same spectrum [3–5].

L. Zhang et al., *AI-enabled Spectrum Sharing*, SpringerBriefs in Computer Science, https://doi.org/10.1007/978-981-97-7644-3_1

In the wireless edge network, two representative technologies for spectrum sharing are *cognitive radio* (CR) [6, 7] and *artificial intelligence* (AI) [8, 9]. Specifically, the CR-enabled spectrum sharing emphasizes the capture of real-time radio environment states and makes spectrum sharing decisions among different wireless links based on the states and highly efficient spectrum sharing models. Different from the CR-enabled spectrum sharing, the AI-enabled spectrum sharing focuses on the statistical pattern of multi-dimensional radio environment variations, and learns directly long-term spectrum sharing policies from historical data. In the following, we first elaborate the main idea of CR and AI to achieve spectrum sharing, and then focus on the preliminaries of AI-enabled spectrum sharing.

1.1.1 CR-Enabled Spectrum Sharing

In a CR network, there are typically two kinds of wireless links, i.e., primary link and secondary link [10–15]. In particular, the primary link has a higher priority to use the spectrum while the priority of the secondary link is lower. CR aims to achieve the spectrum sharing between the primary and secondary links based on their priorities to use the spectrum resource. According to whether the secondary link occupies the spectrum at the same time as the primary link, the spectrum sharing in a CR network can be divided into overlay spectrum sharing and underlay spectrum sharing.

In the overlay spectrum sharing, the secondary link can use the spectrum only if the spectrum is idle, which means that the primary link does not occupy the spectrum, and has to withdraw from the spectrum if the primary link re-occupies the spectrum. In this way, the secondary link will not cause any interference to the primary link and can enhance the spectrum utilization efficiency. An example of overlay spectrum sharing in CR is illustrated in Fig. 1.1, in which the secondary link uses the idle time-frequency resource block for wireless transmissions. Obviously, it is vital to detect correctly whether the spectrum is idle or busy for the secondary link to achieve the overlay spectrum sharing [16, 17]. Otherwise, false detections may cause interference from the secondary link to the primary link, or fail to fully make use of the spectrum resource. For this purpose, spectrum sensing is widely investigated. Existing spectrum sensing techniques include energy detection [18, 19], matched filtering [20, 21], cyclostationary-based sensing [22], etc. Due to the disturbance of time-varying wireless channels and noise, imperfect spectrum sensing is unavoidable. To tackle with this issue, cooperative spectrum sensing among multiple secondary nodes is suggested [23].

In the underlay spectrum sharing, the secondary link is allowed to use the spectrum at the same time as the primary link to further enhance the spectrum utilization efficiency [24, 25]. Obviously, the underlay spectrum sharing will cause mutual interference between secondary link and primary link. An example of the underlay spectrum sharing in CR is illustrated in Fig. 1.2. Since the primary link has the priority to use the spectrum, the secondary link needs to be designed carefully

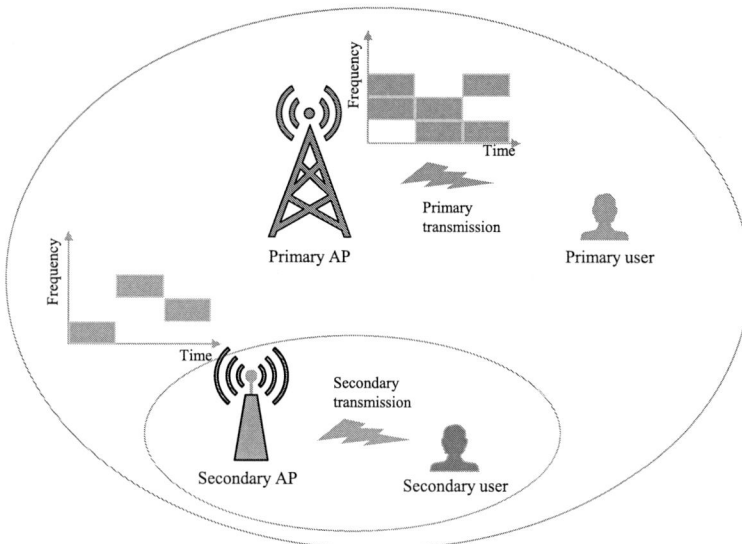

Fig. 1.1 Overlay spectrum sharing in CR

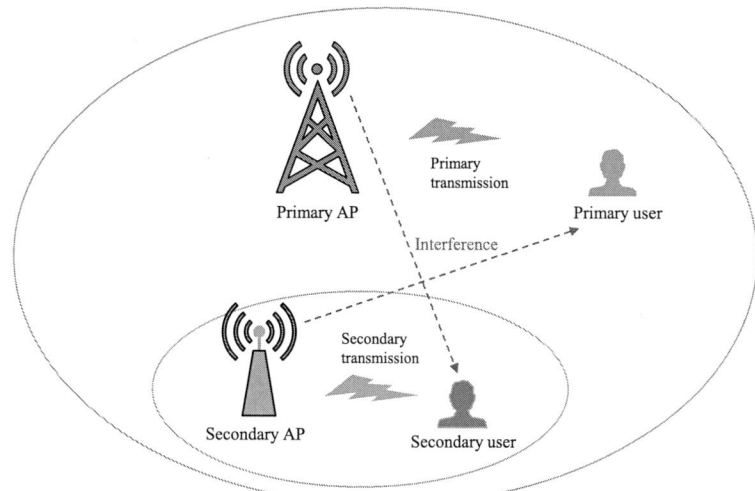

Fig. 1.2 Underlay spectrum sharing in CR

to manage the interference to the primary link and avoid severe performance degradation of the primary transmission. Typically, there are two kinds of designs to manage the interference and guarantee the primary transmission performance. One design is to constrain the received interference power at the primary receiver. Since the interference power at the primary receiver is directly determined by the emitted power of the secondary transmitter and the interference channel quality between the secondary transmitter and the primary receiver. Thus, the interference power at the primary receiver can be well controlled with the information of the interference channel quality [26, 27]. The other design is to guarantee the received primary signal quality at the primary receiver [28, 29], e.g., *signal-to-noise ratio* (SNR) and outage. It is known that the received primary signal quality is highly related to the primary and secondary transmission configurations, the primary channel quality between the primary transmitter and primary receiver, and the interference channel quality. Thus, the primary and secondary transmission configurations can be jointly optimized with the information of primary and interference channel quality, so as to guarantee the received primary signal quality at the primary receiver.

1.1.2 AI-Enabled Spectrum Sharing

AI originates from computer and control areas, and fuels the fast developments of the spectrum sharing in wireless edge networks in the last decades. The AI-enabled spectrum sharing aims to learn long-term spectrum sharing policies, which in fact are the mappings from multi-dimensional radio environment to robust spectrum sharing decisions, from massive historical data in wireless edge networks [30]. According to whether the learning is offline or online, AI-enable spectrum sharing can be divided into *deep learning* (DL) based spectrum sharing [31] and *reinforcement learning* (RL) based spectrum [32]. Extensive researches have been conducted to explore the potential applications and benefits of the deep/reinforcement learning for spectrum sharing to varieties of wireless edge networks. Since the requirements or constraints in various wireless edge networks are quite different, the designs of deep/reinforcement learning in various scenarios are distinct.

Deep learning based spectrum sharing represents a spectrum sharing policy with a *deep neural network* (DNN) as shown in Fig. 1.3, whose structure and parameters determine the corresponding policy. The DNN takes the multi-dimensional radio environment data as the inputs, and outputs the spectrum sharing decision, for instance, the transmission configurations of different wireless links. Deep learning based spectrum sharing can be segmented into two phases, i.e., offline learning phase and online transmission phase [33, 34].

- In the offline learning stage, historical radio environment data needs to be collected and labeled with the target spectrum sharing decisions, which are usually obtained with perfect spectrum sharing models. Then, supervised training is typically adopted to continuously update the DNN parameters with the historical radio environment data as inputs to minimize the error between the DNN outputs and the target spectrum sharing decisions. When the error does not diminish any longer, the DNN parameters converge and the DNN is thought to be well trained.
- In the online transmission phase, the well-trained DNN are deployed to output high-quality spectrum sharing decisions by feeding instantaneous radio environ-

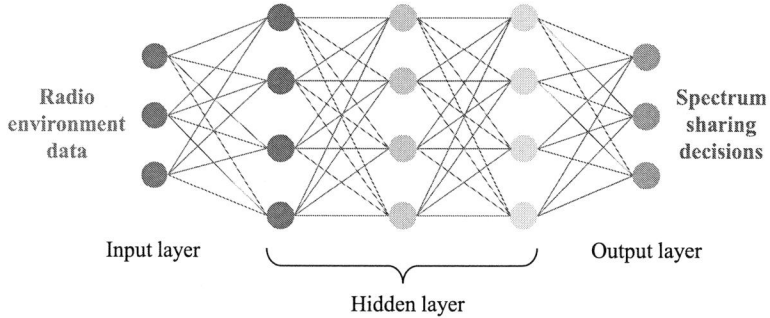

Fig. 1.3 An example of DNN representing the spectrum sharing policy

Fig. 1.4 Reinforcement learning based spectrum sharing policy

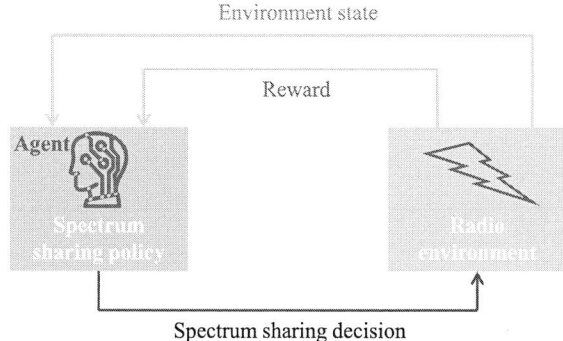

ment data into the DNN. In the quasi-static pr slow-varying radio environment, the well-trained DNN can always output a spectrum sharing policy with good performance, even the DNN has never seen some new radio environment data. Nevertheless, if the radio environment changes dramatically and the distribution of the radio environment data in the online transmission phase is too far away from that in the offline learning phase, it is hard for the DNN to make a good spectrum sharing decision.

Reinforcement learning (RL) based spectrum sharing represents a spectrum sharing policy with a Q-table or a DNN, whose difference will be elaborated in the next part. Different from the deep learning based spectrum sharing, the RL based spectrum sharing learns the spectrum sharing policy in an online manner and makes decisions based on continuous-varying radio environment data [35–37]. In particular, to learn a good spectrum sharing policy, a reward mechanism is established as shown in Fig. 1.4. First, by observing the instantaneous radio environment state, the RL agent makes a spectrum sharing decision, which is used to configure the transmissions of wireless links. Then, the agent receives a reward, i.e., the target of spectrum sharing, for instance, the throughput or energy efficiency of the considered wireless edge network, in terms of the spectrum sharing decisions. Based on the reward, the agent updates its spectrum sharing policy toward the direction of maximizing the long-term reward. In this way, the agent can iteratively learn to update the spectrum sharing policy in a trial and error manner, until the expected long-term reward converges. It should be pointed out that, due to the online learning nature, the RL based spectrum sharing tracks the variations of radio environment and can always make a satisfactory spectrum sharing decision, even the radio environment changes dramatically. This Springer Brief focuses on the RL based spectrum sharing and will elaborate typical RL algorithms as preliminaries of the AI-enabled spectrum sharing in the following chapters.

1.2 Preliminaries

In this part, we will introduce the core AI algorithms that will be used in the subsequent chapters, including general RL, *deep Q-network* (DQN) , and *deep deterministic policy gradient* (DDPG).

1.2.1 General RL

RL enables an agent to learn optimal behavior strategies through exploration and exploitation in an environment. During the RL process, the agent selects actions based on the current state and receives rewards or feedback from the environment, which it uses to adjust its behavior strategy to maximize cumulative rewards over the long term [38, 39].

In recent years, RL has demonstrated significant potential and value in various fields such as autonomous driving, investment decision-making, robotic control, and intelligent gaming. In these applications, RL enables agents to adjust their behavior according to dynamic changes in the environment, allowing them to make optimal decisions in complex scenarios. The core problem in RL research is how to select an action in the current state to maximize reward returns. For example, in the game of Go, the player acts as the agent in RL, while the board represents the environment. The player chooses a move based on the current state of the game, thereby executing an action, and the state of the board changes accordingly. The player's objective is to win the game through a series of actions, and the environment provides rewards based on the outcome.

In general, a RL framework includes six fundamental elements as follows:

1. **Action space** A: the action space is defined as a set consisting of the available actions a at the agent;
2. **State space** S: the state space is defined as a set consisting of the possible states s of the environment;

3. **Immediate reward** $r(s, a)$: the immediate reward $r(s, a)$ is defined as the revenue of selecting action a for the state s;
4. **Transition probability space** P: the transition probability space is defined as a set of transition probabilities $p_{ss'}(a) \in P$ from an environmental state s to an environmental state s' after selecting the action a for the state s;
5. **Policy** $\pi(s) \in A$: the policy is defined as an action selection strategy for the state s.
6. **State-action-value function** $q_{\pi}(s, a)$: the state-action-value function is defined as the long-term expected discounted reward in the future after by selecting a as the action for the state under policy π.

Note that, the transition probability of the environmental state is jointly impacted by the environment and the RL agent's action, and is typically modeled as a *markov decision process* (MDP). The objective of the RL agent maximize the long-term reward instead of the immediate reward by optimizing the action policy. In fact, the long-term reward maximization is not equivalent to the immediate reward maximization. In particular, the long-term reward is related to both its immediate reward and the future rewards, and can be expressed in a recursive form as follows:

$$q_{\pi}(s, a) = r(s, a) + \eta \sum_{s' \in S} \sum_{a' \in A} p_{ss'}(a) q_{\pi}(s', a'), \tag{1.1}$$

where $\eta \in [0, 1]$ is the discount factor. The design objective is typically finding the optimal policy $\pi^*(s)$ subject to the maximal long-term reward $q_{\pi}(s, a)$ under any circumstance or environmental state s. By denoting $q_{\pi^*}(s, a)$ as the largest long-term reward, we can re-express (1.1) as

$$q_{\pi^*}(s, a) = r(s, a) + \eta \sum_{s' \in S} p_{ss'}(a) \max_{a' \in A} q_{\pi^*}(s', a'). \tag{1.2}$$

Then, we can obtain the optimal policy as

$$\pi^*(s) = \arg\max_{a \in A} [q_{\pi^*}(s, a)], \ \forall s \in S. \tag{1.3}$$

1.2.2 DQN

It is demanding to derive the optimal $q_{\pi^*}(s, a)$ from (1.2) in the absence of the transition probability $p_{ss'}(a)$, which is the typical case in practical situations. Then, the Q-learning algorithm is to promising alternative [39]. By constructing a $|S| \times |A|$ Q-table with Q-value element $Q(s, a)$, the policy can be represented by the Q-table. Then, the RL agent can optimize the policy by continuously updating each element

$Q(s, a)$ in the Q-table until convergence. The corresponding element update rule is

$$Q(s, a) \leftarrow (1-\alpha)\, Q(s, a) + \alpha \left[r(s, a) + \eta \max_{a' \in A} Q(s', a') \right], \tag{1.4}$$

where α is the learning rate.

However, the Q-learning algorithm suffers from the curse of dimensionality with an infinitely large state-action space. Then, the DQN is proposed by using a DNN to represent the Q-table in the Q-learning algorithm [40, 41], i.e.,

$$Q(s, a; \theta) \approx \hat{Q}(s, a), \tag{1.5}$$

where θ is the parameter set of the DNN.

Then, by defining the experience of the DRL agent as $e = \langle s, a, r, s' \rangle$, the loss function is introduced to quantify this discrepancy and adjust the parameters of the neural network to minimize the loss function. The loss function can be expressed as

$$\mathbb{L}(\theta) = [y^{\text{Tar}} - Q(s, a; \theta)]^2, \tag{1.6}$$

where y^{Tar} is the target output of the DQN, i.e.,

$$y^{\text{Tar}} = r + \eta \max_{a' \in A} Q(s', a'; \theta). \tag{1.7}$$

Unlike traditional supervised learning, a significant challenge in RL is the delay in reward labels, meaning there is a time gap between the agent's actions and the eventual rewards received. This delay complicates the application of training methods in RL. To address this issue, the DeepMind team introduced the experience replay mechanism. This mechanism disrupts the temporal correlation between samples by maintaining a replay buffer that stores past transitions (combinations of states, actions, and rewards). During training, the neural network randomly samples from this buffer for learning. The experience replay mechanism not only mitigates the delayed reward problem in RL but also enables the agent to learn from historical experiences rather than relying solely on the most recent data. This ability allows the DQN algorithm to excel in handling problems with complex delayed reward structures, further enhancing the algorithm's performance and applicability.

1.2.3 DDPG

The Q-learning and the DQN algorithm belong to the class of value-based RL methods. Their basic ideas both lie in evaluating the value of each action and selecting the optimal action based on these evaluations. It should be noted that in such a method, the long-term reward needs to be estimated for each state-action pair to obtain the optimal policy, and when the action space is continuous, these two algorithms are inapplicable. In contrast to value-based methods, *policy gradient* (PG) methods directly optimize the policy. In PG methods, the policy is represented as a mapping from states to actions, typically implemented through a parameterized function such as a neural network. The updates in PG algorithms are usually based on the principle of gradient ascent, where the gradient of the policy concerning its parameters is calculated, and the parameters are updated in the direction of this gradient.

To implement PG algorithms, the Actor-Critic architecture is a common framework. Compared to traditional policy gradient methods, the Actor-Critic architecture combines two main components: the Actor and the Critic. The Actor is responsible for generating actions and corresponds directly to the policy network, selecting actions based on the current policy. The Critic, on the other hand, evaluates the performance of the current policy, typically by approximating the Q-value function or other value functions. This structure allows the Actor-Critic algorithm to update at each step, rather than waiting for the entire episode to conclude as in traditional policy gradient methods. Because the Critic network provides immediate feedback, the learning process of the Actor-Critic algorithm is more stable and can adapt more quickly to changes in the environment. This design enables the Actor-Critic algorithm to excel in many complex RL problems, making it an important branch in the current field of RL.

Based on the ideas of DQN and the Actor-Critic framework, Lillicrap et al. improved the *deterministic policy gradient* (DPG) method proposed in [42], and developed a DDPG in [43]. DDPG employs two deep neural networks to represent the policy function and the value function, i.e., an actor DNN $\mu(s; \theta_\mu)$ and a critic DNN $Q(s, a; \theta_Q)$ respectively. In particular, the actor DNN $\mu(s; \theta_\mu)$ is the policy network and is used to calculate a deterministic and continuous action a in state s, and the critic DNN is the value-function network and is used to estimate the long-term reward by executing the action a in state s. Besides, a target actor DNN $\mu^-(s; \theta_\mu^-)$ and a target critic DNN $Q^-(s, a; \theta_Q^-)$ are established to stabilize the training of actor DNN and critic DNN.

To train the actor DNN and critic DNN, the DDPG agent randomly samples batches of experiences from the experience replay buffer similar to the DQN. Differently, the DDPG accumulates experiences in a unique way compared with the DQN. In particular, the executed action is the addition of the noisy and the actor DNN's output, i.e., $a = \mu(s; \theta_\mu) + \zeta$, where ζ is a random variable and is used to continuously exploring the action space. This is a major difference compared with the DQN, which typically adopts the epsilon-greedy method for discrete action explorations. Existing literature demonstrates that adding noise is a preferable method to explore continuous action space [43].

The training of DDPG aims to maximize the expected cumulative discounted reward. Specifically, its goal is to maximize the following expected value:

$$J\left(\theta_\mu\right) = \mathrm{E}_{\theta_\mu}\left[r_1 + \gamma r_2 + \gamma^2 r_3 + \cdots\right]. \tag{1.8}$$

Next, the objective function is optimized using the method of gradient ascent. The gradient of the objective function with respect to θ^{μ} is:

$$\frac{\partial J\left(\theta_{\mu}\right)}{\partial \theta_{\mu}} = \mathrm{E}_s\left[\frac{\partial Q\left(s, a; \theta_Q\right)}{\partial \theta_{\mu}}\right]. \tag{1.9}$$

Given the deterministic policy $a = \mu\left(s; \theta_{\mu}\right)$, it follows that:

$$\frac{\partial J\left(\theta_{\mu}\right)}{\partial \theta_{\mu}} = \mathrm{E}_s\left[\frac{\partial Q\left(s, a; \theta_Q\right)}{\partial a}\frac{\partial \mu\left(s; \theta_{\mu}\right)}{\partial \theta_{\mu}}\right]. \tag{1.10}$$

The training procedure of the critic DNN is similar to that of DQN: by sampling a mini-batch of experiences from the experience replay buffer D, the DDPG agent can update θ_Q by adopting a proper optimizer (e.g., Adam and SGD) method to minimize the expected prediction error of the sampled experiences, with the gradient given by:

$$\frac{\partial L\left(\theta_Q\right)}{\partial \theta_Q} = \mathrm{E}_{s,a,r,s'\sim D}\left[\left(y^{\mathrm{Tar}} - Q\left(s, a; \theta_Q\right)\right)\frac{\partial Q\left(s, a; \theta_Q\right)}{\partial \theta_Q}\right], \tag{1.11}$$

where the target value function y^{Tar} is calculated by combining the immediate reward with the expected Q-value of the next state, given by

$$y^{\mathrm{Tar}} = r + \gamma Q^-\left(s', \mu^-\left(s'; \theta_{\mu}^-\right); \theta_Q^-\right), \tag{1.12}$$

where θ_{μ}^- and θ_Q^- represent the parameters of the target policy network and target value network, respectively. The parameters of the target networks are periodically updated based on the current network parameters, i.e.,

$$\theta_{\mu}^- \leftarrow \tau\theta_{\mu} + (1 - \tau)\theta_{\mu}^-,$$
$$\theta_Q^- \leftarrow \tau\theta_Q + (1 - \tau)\theta_Q^-. \tag{1.13}$$

By computing the gradient of the Q-value function with respect to the action, the Critic network provides feedback on how to improve the policy. The Actor network then updates its parameters in the direction that increases the Q-value, specifically updating in the direction that maximizes the expected Q-value. This process can be realized using gradient ascent, with the specific gradient update formula as follows:

$$\theta_{\mu} \leftarrow \theta_{\mu} + \alpha\nabla_{\theta_{\mu}}Q\left(s, \mu\left(s; \theta_{\mu}\right); \theta_Q\right), \tag{1.14}$$

where α is the learning rate. Through this approach, the Actor network learns to select better actions, thereby improving overall performance.

1.3 Structure of the Brief

In this Brief, we apply the RL technique in AI to spectrum sharing in wireless edge networks. In particular, we focus on opportunistic spectrum sharing scenario, centralized spectrum sharing scenario, and distributed spectrum sharing scenario. To enhance the spectrum efficiencies in the three scenarios, we design novel transmission strategies based on the RL and conduct extensive evaluations.

In Chap. 2, we consider an opportunistic spectrum sharing scenario, in which secondary links are allowed to use the spectrum resource when it is idle or not occupied by primary links. For this purpose, spectrum sensing is adopted by secondary nodes to detect the idle/occupied state of the spectrum resource. Due to the imperfect spectrum sensing caused by the wireless channel and noise, the secondary transmissions may collide with the primary transmissions and cause severe interference to degrade the performance, e.g., outage, of the primary links. To protect primary transmissions, the DQN-based *modulation and coding scheme* (MCS) selection method is designed for the primary transmitter. With the method, the primary transmitter is able to predict the received interference in the future and selects a proper MCS in advance to enhance the transmission performance.

In Chap. 3, we consider a centralized spectrum sharing scenario, in which multiple *base stations* (BSs) use the same spectrum resource to serve the associated users simultaneously and all the BS-User links suffer from severe interference. To manage the interference, a centralized power control method is developed for all the BSs. In particular, all the BSs aggregate their historical local wireless data into the cloud, which adopts a DDPG based centralized learning mechanism to construct a power control policy for each BS. With the learned policy, all BSs can configure their local transmit power to enhance the network spectrum efficiency in the absence of simultaneous information exchange among neighbors.

In Chap. 4, we consider a distributed spectrum sharing scenario, in which multiple BSs use the same spectrum resource to serve the associated users simultaneously and all the BS-User links suffer from severe interference. To manage the interference, a distributed power control method is developed for all the BSs. In particular, each BS learns a distributed power control policy by interacting with the environment in an online manner. A cloud-assisted reward mechanism is proposed to accelerate the learning. With the learned policy, each BS can adapt its transmit power independently to enhance the network spectrum efficiency.

In Chap. 5, we conclude this Brief and provide potential research directions toward the spectrum sharing in wireless edge networks.

References

1. Tariq, F., Khandaker, M.R.A., Wong, K.-K., Imran, M.A., Bennis, M., Debbah, M.: A speculative study on 6G. IEEE Wirel. Commun. **27**(4), 118–125 (2020)
2. Alsaedi, W.K., Ahmadi, H., Khan, Z., Grace, D.: Spectrum options and allocations for 6G: A regulatory and standardization review. IEEE Open J. Commun. Soc. **4**, 1787–1812 (2023)

3. Zhang, L., Liang, Y.-C., Xiao, M.: Spectrum sharing for Internet of Things: a survey. IEEE Wirel. Commun. **26**(3), 132–139 (2019)
4. Matinmikko-Blue, M., Yrjölä, S., Ahokangas, P.: Spectrum management in the 6G era: the role of regulation and spectrum sharing. In: Proceedings of the 2nd 6G Wireless Summit (6G SUMMIT), pp. 1–5 (2020)
5. FCC Technology Advisory Council: 6G Working Group Position Paper, Aug. 2023 [Online]. Available: https://www.fcc.gov/sites/default/files/Consolidated_6G_Paper_FCCTAC23_Final_for_Web.pdf
6. Tsiropoulos, G.I., Dobre, O.A., Ahmed, M.H., Baddour, K.E.: Radio resource allocation techniques for efficient spectrum access in cognitive radio networks. IEEE Commun. Surveys Tuts. **18**(1), 824–847 (2016)
7. Yang, C., Li, J., Guizani, M., Anpalagan, A., Elkashlan, M.: Advanced spectrum sharing in 5G cognitive heterogeneous networks. IEEE Wirel. Commun. **23**(2), 94–101 (2016)
8. Dang, L., Wang, W., Tse, C.K., Lau, F.C.M., Wang, S.: Smooth deep reinforcement learning for power control for spectrum sharing in cognitive radios. IEEE Trans. Wirel. Commun. **21**(12), 10621–10632 (2022)
9. Doshi, A., Yerramalli, S., Ferrari, L., Yoo, T., Andrews, J.G.: A deep reinforcement learning framework for contention-based spectrum sharing. IEEE J. Sel. Areas Commun. **39**(8), 2526–2540 (2021)
10. Haykin, S.: Cognitive radio: Brain-empowered wireless communications. IEEE J. Select. Areas Commun. **23**(2), 201–220 (2005)
11. ElSawy, H., Hossain, E., Kim, D.I.: HetNets with cognitive small cells: user offloading and distributed channel allocation techniques. IEEE Commun. Mag. **51**(6), 28–36 (2013)
12. Andrews, J.G., Baccelli, F., Ganti, R.K.: A tractable approach to coverage and rate in cellular networks. IEEE Trans. Commun. **59**(11), 3122–3134 (2011)
13. Cheung, W., Quek, T.Q.S., Kountouris, M.: Throughput optimization, spectrum allocation, access control in two-tier femtocell networks. IEEE J. Sel. Areas Commun. **30**(3), 561–574 (2012)
14. Mukherjee, S.: Distribution of downlink SINR in heterogeneous cellular networks. IEEE J. Sel. Areas Commun. **30**(3), 575–585 (2012)
15. Lien, S.-Y., Chen, K.-C., Liang, Y.-C., Lin, Y.: Cognitive radio resource management for future cellular networks. IEEE Wirel. Commun. **21**(1), 70–79 (2014)
16. Liang, Y.-C., Zeng, Y., Peh, E.C.Y., Hoang, A.T.: Sensing-throughput tradeoff for cognitive radio networks. IEEE Trans. Wirel. Commun. **7**(4), 1326–1337 (2008)
17. Zhang, L., Xiao, M., Wu, G., Alam, M., Liang, Y.-C., Li, S.: A survey of advanced techniques for spectrum sharing in 5G networks. IEEE Wirel. Commun. **24**(5), 44–51 (2017)
18. Digham, F.F., Alouini, M.S., Simon, M.K.: On the energy detection of unknown signals over fading channels. IEEE Trans. Commun. **55**(1), 21–24 (2007)
19. Herath, S.P., Rajatheva, N., Tellambura, C.: Energy detection of unknown signals in fading and diversity reception. IEEE Trans. Commun. **59**(9), 2443–2453 (2011)
20. Ma, L., Li, Y., Demir, A.: Matched filtering assisted energy detection for sensing weak primary user signals. IEEE International Conference on Acoustics, Speech, and Signal Processing(ICASSP) (2012)
21. Zhang, X., Chai, R., Gao, F.: Matched filter based spectrum sensing and power level detection for cognitive radio network. IEEE Global Conference on Signal and Information Processing (Global SIP) (2014)
22. Bogale, T.E., Vandendorpe, L.: Multi-cycle cyclostationary based spectrum sensing algorithm for OFDM signals with noise uncertainty in cognitive radio networks. IEEE Military Communications Conference (MILCOM) (2012)
23. Atapattu, S.: Energy Detection based cooperative spectrum sensing in cognitive radio networks. IEEE Trans. Wirel. Commun. **10**(4), 1232–1241 (2011)
24. Liang, Y.C., Chen, K.C., Li, J.Y., Mahonen, P.: Cognitive radio networking and communications: an overview. IEEE Trans. Veh. Technol. **60**(7), 3386–3406 (2011)

25. Jorswieck, E.A., Badia, L., Fahldieck, T., Karipidis, E., Luo, J.: Spectrum sharing improves the network efficiency for cellular operators. IEEE Commun. Mag. **52**(3), 129–136 (2014)
26. Zhang, R.: On active learning and supervised transmission of spectrum sharing based cognitive radios by exploiting hidden primary radio feedback. IEEE Trans. Commun. **58**(10), 2960–2970 (2010)
27. Zhang, L., Xiao, M., Wu, G., Zhao, G., Liang, Y.C., Li, S.: Proactive cross-channel gain estimation for spectrum sharing in cognitive radio. IEEE J. Sel. Areas Commun. **34**(10), 2776–2790 (2016)
28. El Tanab, M., Hamouda, W.: Resource allocation for underlay cognitive radio networks: a survey. IEEE Commun. Surv. Tut. **19**(2), 1249–1276 (2017)
29. Musavian, L., Aïssa, S.: Capacity and power allocation for spectrum-sharing communications in fading channels. IEEE Trans. Wirel. Commun. **8**(1), 148–156 (2009)
30. He, Y., Zhang, Z., Yu, F.R., Zhao, N., Yin, H., Leung, V.C.M., Zhang, Y.: Deep-reinforcement-learning-based optimization for cache-enabled opportunistic interference alignment wireless networks. IEEE Trans. Veh. Technol. **66**(11), 10433–10445 (2017)
31. Bhatti, F.A., Khan, M.J., Selim, A., Paisana, F.: Shared spectrum monitoring using deep learning. IEEE Trans. Cognit. Commun. Network. **7**(4), 1171–1185 (2021)
32. Li, X., Fang, J., Cheng, W., Duan, H., Chen, Z., Li, H.: Intelligent power control for spectrum sharing in cognitive radios: a deep reinforcement learning approach. IEEE Access **6**, 25463–25473 (20180
33. Kumar, R., Singh, C.K., Upadhyay, P.K., Salhab, A.M., Nasir, A.A., Masood, M.: IoT-inspired cooperative spectrum sharing with energy harvesting in UAV-assisted NOMA networks: deep learning assessment. IEEE Internet Things J. **10**(24) (2023)
34. Jiang, F., Chang, D.-W., Ma, S., Hu, Y.-J., Xu, Y.-H.: A residual learning-aided convolutional autoencoder for SCMA. IEEE Commun. Lett. **27**(5) (2023)
35. Dang, L., Wang, W., Tse, C.K., Lau, F.C.M., Wang, S.: Smooth deep reinforcement learning for power control for spectrum sharing in cognitive radios. IEEE Trans. Wirel. Commun. **21**(12) (2022)
36. Doshi, A., Yerramalli, S., Ferrari, L., Yoo, T., Andrews, J.G.: A deep reinforcement learning framework for contention-based spectrum sharing. IEEE J. Sel. Areas Commun. **39**(8) (2021)
37. Foukas, X., Marina, M.K., Kontovasilis, K.: Iris: deep reinforcement learning driven shared spectrum access architecture for indoor neutral-host small cells. IEEE J. Sel. Areas Commun. **37**(8) (2019)
38. Arulkumaran, K., Deisenroth, M.P., Brundage, M., Bharath, A.A.: Deep reinforcement learning: a brief survey. IEEE Trans. Signal Process. **34**(6) (2017)
39. Watkins, C.J., Dayan, P.: Technical note: q-learning. Mach. Learn. **8**, 279–292 (1992)
40. Mnih, V., Kavukcuoglu, K., Silver, D., Graves, A., Antonoglou, I., Wierstra, D., Riedmiller, M.: Playing atari with deep reinforcement learning. Available in arXiv:1312.5602 [cs.LG]
41. Luong, N.C., Hoang, D.T., S., Niyato, D., Wang, P., Liang, Y.-C., Kim, D.I.: Applications of deep reinforcement learning in communications and networking: a survey. Available in arXiv:1810.07862 [cs.NI]
42. Silver, D., Lever, G., Heess, N., Degris, T., Wierstra, D., Riedmiller, M.: Deterministic policy gradient algorithms. Int. Conf. Mach. Learn. **32**(1), 387–395 (2014)
43. Lillicrap, T.P., Hunt, J.J., Pritzel, A., Heess, N., Erez, T., Tassa, Y., Silver, D., Wierstra, D.: Continuous control with deep reinforcement learning. Available in arXiv:1312.5602 [cs.LG]

Chapter 2
AI-Enabled Opportunistic Spectrum Sharing

Abstract This chapter considers the AI-enabled opportunistic spectrum sharing between primary users and secondary users in wireless edge networks, where primary users have priorities to use the spectrum and the secondary users can only access the spectrum when the spectrum is sensed to be idle. However, due to the imperfect spectrum sensing, the secondary users may cause interference to primary users. To deal with this issue, this chapter proposes a DQN based modulation and coding scheme (MCS) selection method for primary users to learn the interference pattern and predict the upcoming interference, which can be used to design proper MCS in advance. Simulation results show that, if the MCS switching is ideal, the primary transmission rate of the proposed algorithm is 90 ~ 100% of the optimal MCS selection method, which knows interference information at primary users as a prior, and the proposed algorithm is 30 and 100% better than the upper confidence bandit (UCB) learning algorithm and the signal-to-noise ratio (SNR) based algorithm in terms of primary transmission rate respectively. If the MCS switching cost is considered, the proposed algorithm still outperforms the benchmark algorithms without increasing system overheads.

Keywords Cognitive HetNet · Intelligent DRL · MCS selection · Spectrum sharing · Switching cost

2.1 Introduction

Fueled by the explosive increase of diversified wireless end devices and wireless service requests, wireless data traffics has continued growing fast in these years and will be a heavy burden for conventional wireless edge networks [1–3]. To accommodate the ever increasing wireless data traffics, it is crucial to enhance the capacity of wireless edge networks. One typical approach is to boost the wireless link efficiency. Unfortunately, the wireless link efficiency has almost approached its theoretical limit with the dozens of year's development of both the multiple-input-multiple-output and orthogonal frequency division multiplexing. As such, the

L. Zhang et al., *AI-enabled Spectrum Sharing*, SpringerBriefs in Computer Science, https://doi.org/10.1007/978-981-97-7644-3_2

spectrum sharing among heterogeneous nodes is emerging as a promising paradigm to improve the network capacity [4–10].

In conventional wireless edge networks, dedicated spectrum resources are allocated to different wireless links, which can occupy the exclusive spectrum at any time. Although dedicated spectrum allocation can guarantee the quality of each single wireless link, it suffers from degraded spectrum efficiency. As the wireless end devices and wireless service requests increase, the spectrum resource shortage issue becomes severe and it is challenging to allocate orthogonal spectrum to different wireless links. In a typical future edge network, wireless nodes with different priorities share the same spectrum. The wireless nodes with the higher priority to use the spectrum are named as primary nodes or users, and the wireless nodes with the lower priority to use the spectrum are named as secondary nodes or users [11–17]. In particular, the primary users can use the spectrum at any time, while secondary users can use the spectrum only if the spectrum is idle to avoid interference to primary users. When the primary users re-occupy the spectrum, secondary users have to withdraw from the spectrum. This spectrum sharing mechanism is called heterogeneous spectrum sharing and can effectively boost the spectrum utilization efficiency.

For the purpose above, secondary users usually adopt spectrum sensing techniques, such as energy detection, to sense the status of the target spectrum, and determine whether to access the target spectrum band or not. In particular, the secondary transmitter first receives the primary signal to measure the energy. If the detected energy of the received primary signal is higher than a preset threshold, it is probable that the target spectrum band is being occupied by primary transmissions. Accordingly, secondary users keep silent to protect the primary transmissions. Otherwise, the target spectrum band is thought to be idle and secondary users access the target spectrum for secondary transmissions. However, the spectrum sensing may be imperfect due to the disturbance of noise and the deep fading in wireless channels. This may lead to interference from secondary users to primary users when primary users are working and the false detection occurs.

In fact, [16] has taken the imperfect spectrum sensing issue into consideration in early cognitive radio networks, especially when the received primary signal strength at secondary users is relatively low, e.g., the SNR is as low as $-15\,\text{dB}$. In particular, the authors therein suggested to design a high detection probability, e.g., 90%, to reduce the interference from secondary users to primary users due to imperfect spectrum sensing. This method has been demonstrated to be able to reduce effectively the impact of imperfect spectrum sensing on the primary transmission performance and thus is widely adopted in early cognitive networks [3]. However, it is known that the performance of the method in [16] degrades when the primary users, especially the *primary transmitter* (PT), are transparent to secondary users, i.e., the strength of the received primary signals is extremely low (much lower than $-15\,\text{dB}$). In this case, if the interfering channel from the secondary transmitter to the primary receiver is strong, the secondary transmitter may cause severe interference to degrade the primary transmission performance, even leading to outages.

We notice that, due to the spectrum sensing operation at secondary users, the secondary transmission begins later than the primary transmission. Therefore, the interference signal strength from the *secondary transmitter* (ST) is unknown at the primary users when the primary transmission begins. In other words, the primary users cannot adapt their transmission to the interference. Typically, the received interference at the primary receiver is highly related to the secondary transmit power and the interference channel, both of which are subject to certain patterns, for example, the secondary transmit power is constant and the magnitude of the interference channel follows a Rayleigh distribution. Thus, although the primary users cannot adapt their transmissions according to the real-time interference from the secondary transmitter, the primary users can learn the pattern of the received interference by digging the historical interference information and predict the upcoming interference in the subsequent frames. In this chapter, we adopt the DQN algorithm for the primary users to learn the pattern of the received interference and infer the future interference [18]. With the inferred interference, the primary users can intelligently select the optimal modulation modes to boost the transmission rate of primary links.

2.2 Syetem Model

Figure 2.1 depicts the considered heterogeneous spectrum sharing scenario, in which a pair of primary users are conducting uplink transmissions from PT to BS and all secondary users share the same spectrum band for opportunistic secondary transmissions in an overlay mode. To avoid severe co-channel interference from secondary users to the primary transmission, each ST (namely, ST-$k, k \in \{1, 2, \ldots, K\}$) adopts an energy-detection based spectrum sensing approach to identify the state of the spectrum band and determine whether to access it. In the following, we provide the spectrum sharing model, channel and signal models in the considered scenario.

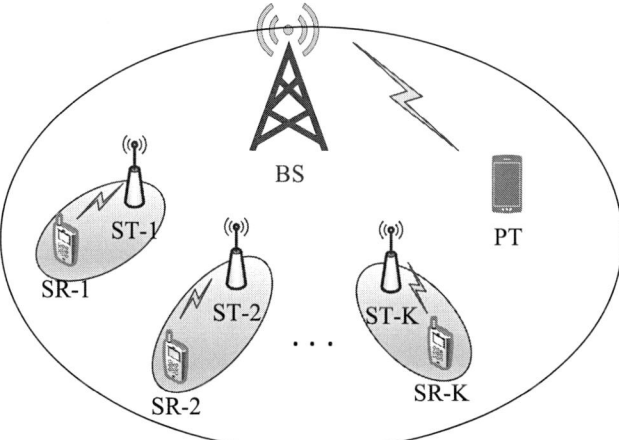

Fig. 2.1 Considered heterogeneous spectrum sharing scenario, which consists of a PT, a BS, and K pairs of ST-SR

2.2.1 Spectrum Sharing Frame Structures

Figure 2.2 gives the frame structures of primary and secondary users for the spectrum sharing. In particular, Fig. 2.2a shows the frame structure of primary transmission, whose duration is T and segmented into an MCS selection phase of duration τ_p and a data transmission phase of duration $T - \tau_p$. Note that τ_p is quite small compared with T in practical situations. In the former phase, the PT transmits pilots to the BS. On the one hand, by receiving the pilots signals at the BS, the channel can be estimated. On the other hand, by receiving the pilot signals at the BS, the SNR can be measured, which can be further used to select the optimal MCS scheme for the primary transmissions. In the data transmission phase, the selected MCS scheme is applied for the uplink data transmission and the BS uses the estimated channel information to equalize the impacts of channels on the data for data recovery.

As aforementioned above, to avoid interference to the primary transmissions, secondary users adopt an energy-detector based spectrum sensing approach to determine whether the spectrum is occupied by the primary users or not [19]. Figure 2.2b shows the frame structure of the secondary transmissions, which is synchronous with the frame structure of primary transmissions. In particular, the frame structure of secondary transmissions consists of two successive phases including spectrum sensing phase with duration τ and data transmission phase with duration $T - \tau$. In the spectrum sensing phase, secondary users detect the energy of primary signals and determine whether the detected energy exceeds a preset threshold or not. If the detected energy does not exceed the preset threshold, the spectrum band is thought to be idle, secondary users will access the spectrum band. Otherwise, secondary users will keep silent.

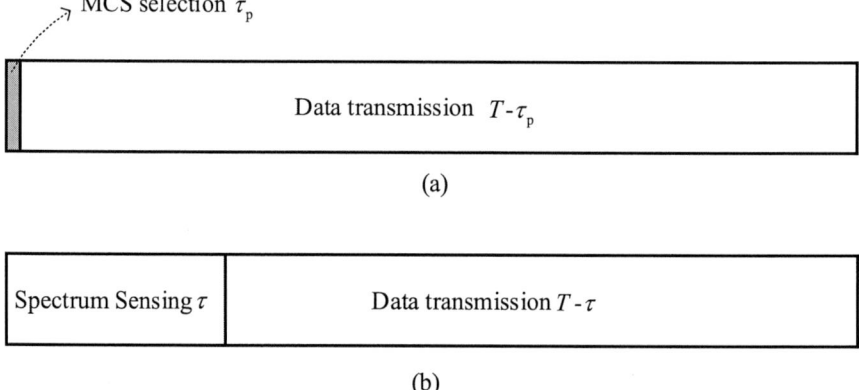

Fig. 2.2 Spectrum sharing frame structures: (**a**) primary transmission frame structure; (**b**) secondary transmission frame structure

In the spectrum sensing stage, i.e., the former τ of each frame, secondary users only detect the energy of primary signals and will not cause any interference to the primary transmission. Due to the imperfect spectrum sensing, secondary users may access the spectrum band even if it has been occupied by the primary users. In this case, the secondary users may cause interference to primary transmissions in the remaining time $T - \tau$. Here, we denote α_k as the miss-detection or interference probability that the kth pair of secondary users access the channel and interfere with the primary transmissions. We can calculate the miss-detection/interference probability α_k according to $\alpha_k = 1 - Q\left(\left(\frac{v_k}{\sigma_k^2} - \phi_k - 1\right)\sqrt{\frac{f_{s,k}\tau_k}{2\phi_k+1}}\right)$ [16, 19], where v_k is the preset threshold of energy detection, σ_k^2 is the noise power, ϕ_k is the average SNR value in terms of the primary signals, $f_{s,k}$ is the sampling frequency, τ_k is the duration of the energy detection, and $Q(x)$ is the complementary Gaussian function $Q(x) = \frac{1}{\sqrt{2\pi}}\int_x^\infty e^{-\frac{z^2}{2}}\,dz$.

2.2.2 Channel and Signal Models

Each wireless channel is affected by a large-scale path-loss and a small-scale block Rayleigh fading. In particular, we denote $\bar{g}_p$ and h_p respectively as the large-scale path-loss and the small-scale block Rayleigh fading of the primary channel. Then, the primary channel gain can be denoted as $g_p = \bar{g}_p|h_p|^2$. Similarly, we denote $\bar{g}_k$ and h_k respectively as the large-scale path-loss and the small-scale block Rayleigh fading of the interference channel between ST-k and the BS, and then the corresponding channel gain can be represented by $g_k = \bar{g}_k|h_k|^2$.

The large-scale path-loss is directly determined by the distance and remains constant for static transmitter and receiver, the small-scale block Rayleigh fading is typically a random variable, which changes across different transmission frames while remaining constant in each single transmission frame. Here, we adopt the Jake's model to characterize the variations of small-scale fadings in adjacent frames [20], i.e.,

$$h(t) = \rho h(t - 1) + \delta, \tag{2.1}$$

where ρ denotes the correlation coefficient of the small-scale fadings in two adjacent realizations, δ denotes a random variable with the distribution $\delta \sim \mathcal{CN}(0, 1 - \rho^2)$, and initialization state $h(0)$ is also a random variable and defined by $h(0) \sim \mathcal{CN}(0, 1)$. Especially, for $\rho = 0$, the Jake's model is equivalent to the *independent and identically distributed* (i.i.d) channel model.

According to [23], the adaptive power control adaptation can contribute negligible performance gain in the case with a large number of MCS levels. As a result, we adopt a constant PT transmit power p_p. Since there does not exist any secondary transmission in the former τ of each frame, the SNR measured at the BS can be written as

$$\gamma_0 = \frac{P_p g_p}{\sigma^2}. \tag{2.2}$$

It is clear that active secondary users may interfere with the primary transmission in the secondary data transmission phase. By using S_a to denote the set of active secondary users and using p_k to denote the corresponding transmit power, the *signal to interference and noise ratio* (SINR) measured at the BS can be written as

$$\gamma_1 = \frac{P_p g_p}{\sum_{k \in S_a} p_k g_k + \sigma^2}. \tag{2.3}$$

Furthermore, consider that the bit-interleaver is applied to primary transmission to tackle deep fadings. Then, the average SINR at the BS is

$$\bar{\gamma} = \frac{\tau - \tau_p}{T - \tau_p} \gamma_0 + \frac{T - \tau}{T - \tau_p} \gamma_1. \tag{2.4}$$

2.3   Optimal MCS Selection Mechanism

2.3.1   Basic Principle of the Optimal MCS Selection

In principle, the MCS needs to be properly designed to balance the average transmission rate of the data packet in a frame for a given average SINR. On the one hand, a low-order MCS contributes to a low *symbol error rate* (SER) and a low *packet error rate* (PER), and can improve the transmission reliability and boost the transmission rate. On the other hand, a low-order MCS also means that a symbol carries fewer data bits, and can degrade the transmission rate. Therefore, it is crucial to select the optimal MCS to enhance the transmission rate. It is worth noting that, the MCS selection functionality can be implemented at either the transmitter (i.e., PT) or the receiver (i.e., BS) [23]. Here, we enable the BS to select the MCS since the computing capability of the BS is typically stronger than the PT and the BS can directly measure the SINR and analyze the interference pattern from secondary users.

2.3.2   The Optimal MCS Selection Method and Challenges

Here, we denote the MCS candidate number as M for the primary uplink transmissions from the PT to the BS, and represent them as MCS_m ($m \in \{1, 2, \cdots, M\}$.

For a given SINR, we denote the SER by adopting MCS_m as $f_m(\bar{\gamma})$, and denote the average transmission efficiency by adopting MCS_m as r_m (bits/symbol).

Further, we denote the number of data symbols in each primary packet or frame as N, and define a packet error event when a data symbol is incorrectly decoded at the BS. In this way, we can approximate the packet error rate of a primary transmission as [21]

$$\rho_m(\bar{\gamma}) \approx 1 - (1 - f_m(\bar{\gamma}))^N . \qquad (2.5)$$

Then, the transmission rate (bits/frame) can be written as

$$R_m(\bar{\gamma}) = r_m[1 - \rho_m(\bar{\gamma})]N. \qquad (2.6)$$

The optimal MCS selection mechanism can select the optimal MCS in the MCS selection phase that maximizes the average transmission rate in a frame, i.e.,

$$m^* = \arg \max_{m \in \{1,2,\cdots,M\}} R_m(\bar{\gamma}), \qquad (2.7)$$

which can be solved when $\bar{\gamma}$ is known at the BS. Firstly, the BS calculates $R_m(\bar{\gamma})$, $m \in \{1, 2, \cdots, M\}$ for each MCS candidate. Secondly, the BS identifies the optimal m^* according to the maximum value of $R_{m^*}(\bar{\gamma})$. According to (2.4), $\bar{\gamma}$ is related to γ_0 and γ_1. Specifically, γ_0 is directly measured by the BS. The derivation of γ_1 needs the interference information of secondary transmissions and cannot be complemented due to the time casualty. Therefore, the BS cannot calculate $\bar{\gamma}$ and identify the optimal MCS from the optimization problem (2.7) in the MCS selection phase.

2.4 DQN-Based MCS Selection Algorithm

In this section, we leverage DQN to design an intelligent MCS selection algorithm, such that the BS can select the optimal MCS to boost the primary transmission rate. In the following, we first provide the basic principle and then elaborate algorithm designs.

2.4.1 Basic Principle

As analyzed above, the optimal MCS selection is determined by the value of $\bar{\gamma}$, which can be calculated based on the value of γ_1. However, the value of γ_1 is affected by the interference from the secondary user. Due to the time causality, the BS cannot obtain the value of γ_1 in its MCS selection stage, meaning that it is impractical for the BS to design the optimal MCS selection. In fact, the interference at the BS typically follows a specific pattern. In particular, the interference to the BS is mainly determined by both each ST's transmit power and the channel gain from each ST to the BS, neither of which is available to the BS. This means that the interference pattern is transparent at the BS. Fortunately, the historical interference in past frames can be measured and recorded at the BS. Therefore, the BS can learn and track the variation pattern of the interference from the historical interference data. With the learned knowledge, the BS can infer the upcoming interference and select the optimal MCS in advance to boost the primary transmission rate in each frame.

Basically, the optimal MCS selection can be treated as an optimal decision-making problem. From [22], DQN is a promising theoretical tool to learn the hidden pattern from historical information with DNN and track the optimal strategy gradually. Therefore, it is possible to adopt DQN to learn the interference pattern at the BS, such that the BS can continuously identify the optimal MCS to maximize the primary transmission rate.

2.4.2 Algorithm Designs Without Switching Costs

To map the optimal MCS selection problem into the DQN algorithm, we design the action space, state space, and immediate reward function in the MCS selection problem as follows.

2.4.2.1 Action Space Design

It is clear that the DQN agent at the BS aims to select the optimal MCS for the primary transmission from the PT to the BS at the beginning of each frame. Thus, we design the action space of the DQN agent in each frame to include all the MCS candidates, i.e.,

$$A = \{\text{MCS}_1, \text{MCS}_2, \ldots, \text{MCS}_M\}, \tag{2.8}$$

where MCS_m ($m \in \{1, 2, \ldots, M\}$) defines a specific MCS. By representing $a(t)$ as the selected action at the beginning of frame t, we have $a(t) \in A$.

2.4.2.2 Immediate Reward Function Design

Note that the optimal MCS selection is used to maximize the primary transmission rate. Then, it is reasonable to design the immediate reward of action for the

DQN agent to be the number of corrected transmit data bits. Then, we design the immediate reward function as

$$r(s, a) = \begin{cases} r_m N, & \text{if successful,} \\ 0, & \text{if failed,} \end{cases} \tag{2.9}$$

which will be also used as $r(t)$ when we refer to the immediate reward in frame t.

2.4.2.3   State Space

In general, each component in the state is supposed to offer knowledge for the optimal MCS selection of the DQN agent. The optimal MCS selection in each frame is highly related to three kinds of information as follows.

- Firstly, the optimal MCS selection in each frame is highly affected by the received signal strength from the PT to the BS. In particular, the DQN agent prefers to select a higher order MCS when the received signal strength from the PT is high, and vice versa. As shown in Fig. 2.2, the BS can obtain the SNR in the MCS selection phase. Then, we include the SNR measured at the BS in the state.
- Secondly, the optimal MCS selection in each frame is highly related to the interference strength from secondary users. Note that the BS cannot directly obtain the interference strength in the MCS selection phase of each frame due to the time causality. Then, the interference strength cannot be designed into the state. Alternatively, we can design the state to include some information about the historical interference strength, so that the DQN agent may learn the interference pattern and infer the upcoming interference, and ultimately use the inferred interference for the MCS selection. Thus, we also include both the SNR and the SINR at the BS in the previous Φ frames into the state.
- Thirdly, the optimal MCS is related to the immediate reward. Intuitively, if the above two kinds of information, especially the second one, can be perfectly obtained, the BS can directly determine the optimal MCS. Considering that the second kind of information can only be inferred by the BS. To enhance the inference accuracy, the state is designed to include the action and its immediate reward in the previous Φ frames.

In summary, we design the state in frame t as

$$s(t) = \{a(t - \Phi), r(t - \Phi), \gamma_0(t - \Phi), \bar{\gamma}(t - \Phi), \ldots,$$
$$a(t - 1), r(t - 1), \gamma_0(t - 1), \bar{\gamma}(t - 1), \gamma_0(t)\}. \tag{2.10}$$

Figure 2.3 shows the proposed DQN-based MCS selection algorithm framework, which consists of the MCS selection phase and the data transmission phase. In either phase, we focus on the functionalities of four main modules at the BS, i.e., signal

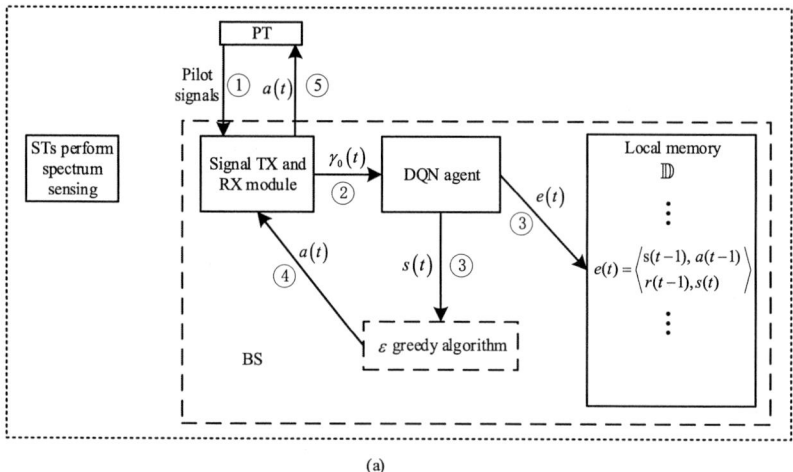

(a)

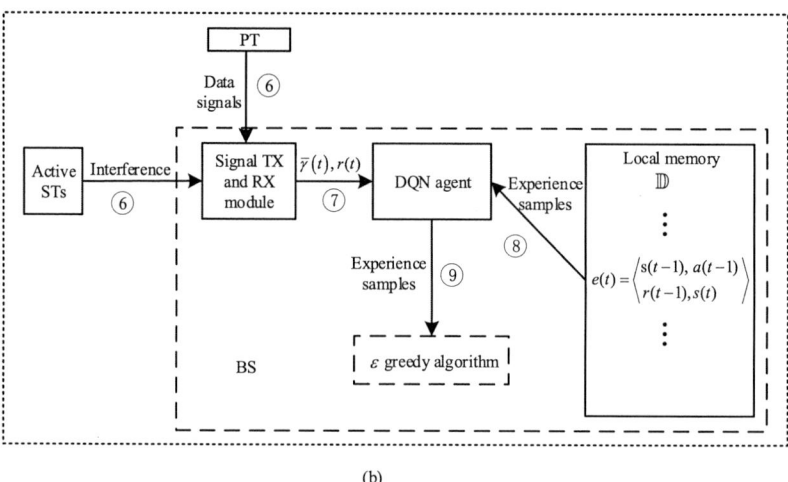

(b)

Fig. 2.3 The proposed DQN-based MCS selection algorithm framework. (**a**) MCS selection phase. (**b**) Data transmission phase

transmitting/receiving module, DQN agent module, local memory $\mathbb{D}$, and ϵ-greedy algorithm module.

In the MCS selection phase, the PT first transmits pilot signals to the signal transmitting/receiving module at the BS. Then, the SNR $\gamma_0(t)$ can be measured in the signal transmitting/receiving module and forwarded to the DQN agent. By fusing the local experiences with SNR $\gamma_0(t)$, the DQN agent can construct the state $s(t) = \{a(t-\Phi), r(t-\Phi), \gamma_0(t-\Phi), \bar{\gamma}(t-\Phi), \ldots, a(t-1), r(t-1), \gamma_0(t-1), \bar{\gamma}(t-1), \gamma_0(t)\}$ and forms a new experience $e(t) = \langle s(t-1), a(t-1), r(t-1), s(t) \rangle$. After that, the DQN agent can store the latest experience into the local memory $\mathbb{D}$, and subsequently input $s(t)$ to the ϵ-greedy algorithm module, which optimizes the selected action $a(t)$ based on $s(t)$ and input it to the signal transmitting/receiving module. Finally, the signal transmitting/receiving module shares the selected $a(t)$ with the PT through a reliable control signaling.

In the data transmission phase, the PT transmits signals to the BS meanwhile active secondary signals cause interference to the BS. Then, the average SINR $\bar{\gamma}(t)$ can be measured in the signal transmitting/receiving module. At the end of this frame, an immediate reward $r(t)$ can be obtained by checking the successfully decoded data bits from the PT. To this end, the DQN agent randomly chooses a mini-batch of experience samples to train the DQN and update the weights and biases therein.

Algorithm 1 shows the pseudocode of the proposed DQN-based MCS selection algorithm. In particular, to stabilize the training of the DQN network and enhance the performance, the "experience replay" and "quasi-static target network" techniques in [22] are adopted. For the experience replay, once constructing a new experience, the DQN agent stores it into the local memory $\mathbb{D}$, whose capacity is N_E experiences, in a first-in-first-out manner. Then, the DQN agent randomly samples a mini-batch of Z experiences from the local memory $\mathbb{D}$ ($Z < N_E$) to train the DQN and update its weights and bias with a certain optimizer instead of training the DQN with all the available N_E experiences. For the quasi-static target network, two DQNs with the same network structure are established, i.e., $\boldsymbol{Q}(s, a; \theta)$ and $\boldsymbol{Q}(s, a; \theta^-)$. In particular, $\boldsymbol{Q}(s, a; \theta)$ is named trained DQN and $\boldsymbol{Q}(s, a; \theta^-)$ is named target DQN. The weights and bias of the trained DQN $\boldsymbol{Q}(s, a; \theta)$ are optimized by experience samples directly and the target DQN is used to replace the trained DQN $\boldsymbol{Q}(s, a; \theta)$ in (1.7). The weights and bias of the target DQN are updated by the weights of the trained DQN periodically, e.g., every L frames. Accordingly, the loss function and the update procedure of θ from (1.6) to (1.7) can be replaced by

$$\mathbb{L}(\theta) = \frac{1}{2Z} \sum_{e \in E} [y^{\text{Tar}} - \boldsymbol{Q}(s, a; \theta)]^2, \tag{2.11}$$

where

$$y^{\text{Tar}} = r + \eta \max_{a' \in A} \boldsymbol{Q}(s', a'; \theta^-), \tag{2.12}$$

$$\theta \leftarrow \theta - \frac{1}{Z} \sum_{e \in E} [y^{\text{Tar}} - \boldsymbol{Q}(s, a; \theta)] \nabla \boldsymbol{Q}(s, a; \theta). \tag{2.13}$$

Algorithm 1 DQN-based MCS selection algorithm.

1: Establish a trained DQN with parameter vector θ and a target DQN with parameter vector θ^-).
2: Initialize randomly the parameter vector θ and define $\theta^- = \theta$.
3: In the t-th frame ($t \leq Z$), the DQN agent randomly selects an MCS as an action to execute, constructs the corresponding experience $\langle s, a, r, s' \rangle$ and stores it in the memory $\mathbb{D}$. Then, the DQN agent can obtain Z experiences after Z frames.
4: **Repeat:**
5: In the t-th frame ($t > Z$), the DQN agent selects an MCS $a(t)$ as an action with the ϵ-greedy policy: select the action $a(t) = \arg\max_{a \in A} \mathbf{Q}(s(t), a; \theta)$ with a probability $1 - \epsilon$, or randomly select an action $a(t)$ from the action space with the probability ϵ.
6: After executing the selected action, the DQN agent can obtain an immediate reward $r(s(t), a(t))$ as designed in (2.9).
7: The DQN agent constructs a new state $s(t + 1)$ in frame $t + 1$.
8: The DQN agent stores the latest experience $\langle s(t), a(t), r(t), s(t + 1) \rangle$ into the local memory $\mathbb{D}$.
9: The DQN agent samples randomly a mini-batch of Z experiences from the local memory $\mathbb{D}$.
10: The DQN agent optimizes the parameter vector θ of the trained DQN with (2.13).
11: For every L frames, use the parameter vector of the trained DQN agent to update that of the target DQN, i.e., $\theta^- = \theta$.

2.4.3 Algorithm Designs with Switching Costs

It is observed that the selected MCS is highly related to the observed state at the DQN agent. In fact, the observed state at the DQN agent may change rapidly due to the fast variations of channel quality, interference strength, and noise. To enhance the transmission performance of the uplink, the DQN agent may switch different optimal MCSs frequently, causing significant system overheads. We also observe that Algorithm 1 may lead to some unnecessary switchings, which have an ignorable impact on the long-term reward. For example, at the beginning of frame $t - 1$, the observed state is $s(t - 1)$ and the DQN can be denoted as $\mathbf{Q}(s, a; \theta(t - 2))$. Then, the selected action is $a(t-1) = \arg\max_{a \in A} \mathbf{Q}(s(t-1), a; \theta(t-2))$ and the updated DQN can be denoted as $\mathbf{Q}(s, a; \theta(t - 1))$. At the beginning of frame t, the observed state is $s(t)$ and the selected action should be $a(t) = \arg\max_{a \in A} \mathbf{Q}(s(t), a; \theta(t - 1))$. If $a(t)$ is distinct from $a(t - 1)$, an MCS switching will happen. However, $\mathbf{Q}(s(t), a(t); \theta(t - 1))$ may be slightly higher than $\mathbf{Q}(s(t), a(t - 1); \theta(t - 1))$ and the MCS switching gain is ignorable from the aspect of the long-term reward. In this case, the MCS switching is unnecessary, and avoiding this switching can reduce system overheads.

To address the issue above, we design a switching cost factor c and refine the immediate reward function as

$$
r(t) = \begin{cases}
r_m N, & \text{if } a(t) = a(t - 1) \text{ and successful,} \\
r_m N - c, & \text{if } a(t) \neq a(t - 1) \text{ and successful,} \\
0, & \text{if } a(t) = a(t - 1) \text{ and failed,} \\
-c, & \text{if } a(t) \neq a(t - 1) \text{ and failed.}
\end{cases}
\tag{2.14}
$$

In particular, c is introduced as a relative value with respect to the transmitted data bits and represents the impact of an MCS switching on the system overhead [23, 24]. If we use (2.14) instead of (2.9) to run Algorithm 1 and adapt the value of c, the DQN agent can effectively strike a balance between the transmission rate and system overheads. In this way, the refinement makes Algorithm 1 more flexible by adjusting the value of the switching cost factor. Simulation results in the next section will demonstrate the effectiveness of the refinement.

2.5   Simulation Results

In this section, we evaluate the performance of the proposed DQN-based MCS selection algorithm through extensive simulations. To demonstrate the advantages, we also provide the performance of the optimal MCS selection algorithm assuming that the BS can obtain the average SINR $\bar{\gamma}$ as a prior. Then, we can solve (2.7) to obtain the optimal solution as the upper bound of the proposed algorithm. Besides, we provide the performance of two benchmark algorithms, including the SNR-based algorithm as well as the upper confidence bandit (UCB) learning algorithm [24, 25]. Specifically, the SNR-based algorithm uses the measured SNR γ_0 to replace the average SINR $\bar{\gamma}$ in (2.7) and obtains a sub-optimal solution by solving (2.7). The UCB learning algorithm selects the optimal MCS as follows:

$$m^* = \arg \max_{m \in \{1,2,\cdots,M\}} \left(\mu_m + \sqrt{\frac{2 \ln t}{\Gamma_m(t-1)}} \right), \qquad (2.15)$$

where $\Gamma_m(t-1)$ is the times that MCS_m has been selected in the previous $t-1$ frames, and μ_m is initialized randomly and updated according to

$$\mu_{m^*} \leftarrow \mu_{m^*} + \frac{1}{\Gamma_{m^*}(t)} \left(r(t) - \mu_{m^*} \right). \qquad (2.16)$$

2.5.1   Assumptions and Settings

Note that the imperfect spectrum sensing will cause interference from STs to primary transmissions only when the primary transmission is active. Here, we evaluate the performance of the proposed algorithm by assuming that the primary transmission is active all the time.

In the simulation, we consider an uncoded primary system in which MCS selection refers to the modulation scheme selection for simplicity. In particular, we consider four MCS candidates as shown in Table 2.1. We implement the DQN with Python whose backend is Tensorflow. In particular, the structure of the DQN

Table 2.1 MCS candidates and the corresponding theoretical SERs in the simulation

MCS	SER [26]
BPSK	$f_1(\bar{\gamma}) = Q\left(\sqrt{2\bar{\gamma}}\right)$
QPSK	$f_2(\bar{\gamma}) = 2\left(1 - \frac{1}{\sqrt{4}}\right) Q\left(\sqrt{\frac{3\log_2(4)\bar{\gamma}}{4-1}}\right)$
16QAM	$f_3(\bar{\gamma}) = 2\left(1 - \frac{1}{\sqrt{16}}\right) Q\left(\sqrt{\frac{3\log_2(16)\bar{\gamma}}{16-1}}\right)$
64QAM	$f_4(\bar{\gamma}) = 2\left(1 - \frac{1}{\sqrt{64}}\right) Q\left(\sqrt{\frac{3\log_2(64)\bar{\gamma}}{64-1}}\right)$

includes one input layer, two hidden layers and an output layer. There are $4\Phi + 1$ neurons with linear activation function in the input layer, each neuron corresponds to one of the $4\Phi + 1$ components in the state $s(t)$. There are 100 neurons with the Relu activation function in each hidden layer. There are four neurons with softmax activation function in the output layer, and each neuron corresponds to one of the four MCS levels in Table 2.1. Besides, we apply an adaptive ϵ-greedy algorithm, in which ϵ follows $\epsilon(t + 1) = \max\{\epsilon_{\min}, (1 - \lambda_\epsilon)\epsilon(t)\}$ [27]. The reason to adopt a decaying ϵ is as follows. In the early learning stage of the proposed algorithm, the captured knowledge of the DQN is quite limited and the DQN agent needs to explore more actions to possibly approach the optimal MCS selection policy and enhance the long-term reward. In the later learning stage of the proposed algorithm, the DQN has obtained enough knowledge about the quasi-optimal even optimal MCS selection policy, and is inclined to exploit the knowledge. We set $\epsilon(0) = 0.3$, $\epsilon_{\min} = 0.005$, and $\lambda_\epsilon = 0.0001$. Besides, the batch size is set to be $Z = 32$ for experience sampling, and the capacity of the local memory at the DQN agent is $N_E = 500$. Furthermore, we set $\eta = 0.5$, and adopt the RMSProp optimizer with a learning rate 0.01 to update θ [28]. In addition, we set $\frac{\tau - \tau_p}{T - \tau_p} = 0.1$ and $\frac{T - \tau}{T - \tau_p} = 0.9$ in (2.4), $L = 100$, and each frame contains $N = 1000$ symbols.

2.5.2   Performance in Quasi-Static and Dynamic Interference Scenarios

In this part, we compare the performance of the proposed algorithm with benchmarks in both quasi-static interference scenario and dynamic-interference scenario.

- *Quasi-static interference scenario.* In the scenario, two pairs of secondary users are considered, denoted by (ST-1, SR-1) and (ST-2, SR-2), and the interference probability is set to be 1, which is the case when the wireless links from the PT to secondary users are completely blocked and secondary users cannot detect PT's signal. Meanwhile, we set the correlation coefficient of each fading channel in two adjacent frames as 0.99. In this scenario, the received interference at the BS varies slowly.

- *Dynamic-interference scenario.* In this scenario, three pairs of secondary users are considered, denoted by (ST-1, SR-1), (ST-2, SR-2), and (ST-3, SR-3). The wireless links from the PT to the former two pairs of secondary users are blocked and the corresponding secondary users cannot detect PT's signal, while the wireless link from the PT to (ST-3, SR-3) is partially blocked and is extremely weak. As such, we set the interference probabilities as 1, 1, and 0.5, respectively. In addition, we consider independent fading channels, i.e., the correlation coefficient of each fading channel is 0. In this scenario, the received interference at the BS changes rapidly.

Figure 2.4 compares the average transmission rates of different algorithms in the quasi-static interference scenario, in which the average SNR $\frac{p_p \bar{g}_p}{\sigma^2}$ measured at the BS is 20 dB and average interference-to-noise ratio $\frac{p_k \bar{g}_k}{\sigma^2}$ ($k \in \{1, 2\}$) measured at the BS is 5 dB. From the figure, we observe that the average transmission rate of the optimal MCS selection fluctuates around 3 kbits/frame, the average transmission rate of the UCB learning algorithm grows from around 1.8 to 2.1 in kbits/frame, and the average transmission rate of the SNR-based algorithm lies between 1 and 1.4 in kbits/frame. Meanwhile, the average transmission rate of the proposed DQN-based MCS selection algorithm gradually approaches the average transmission rate of the optimal MCS selection. Thus, the average transmission rate of the proposed

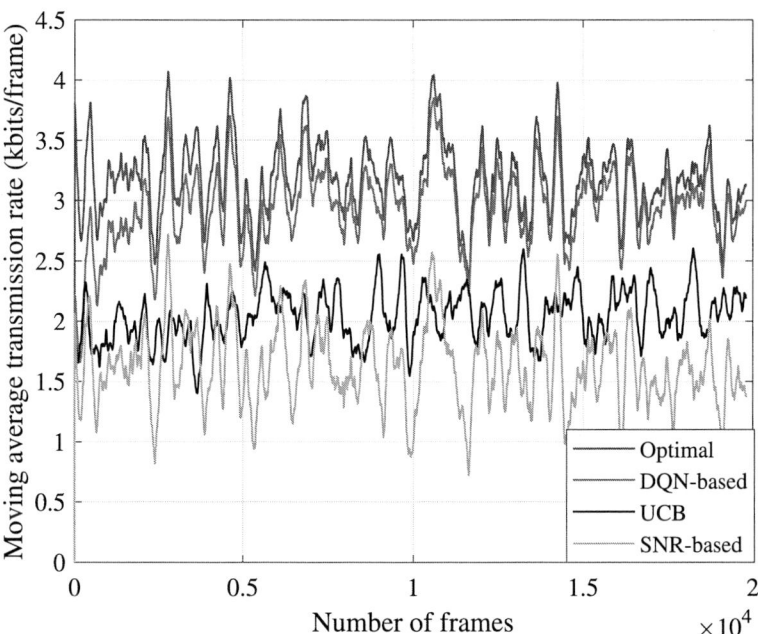

Fig. 2.4 Average transmission rates of different algorithms in the quasi-static interference scenario. Each value is the moving average in the previous 200 frames and each curve is obtained by averaging the data in 20 independent trials

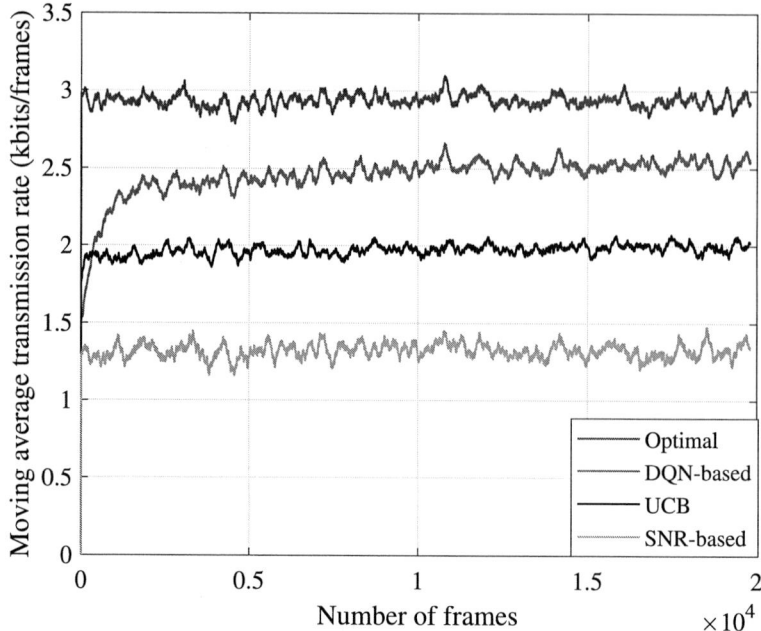

Fig. 2.5 Average transmission rate in the dynamic interference scenario. Each value is a moving average in the previous 200 frames and each curve is obtained by averaging 20 independent trials

DQN-based MCS selection algorithm is around 50% higher than the transmission rate of the UCB learning algorithm, and is 100% higher than the transmission rate of the SNR-based algorithm. This figure indicates that the proposed DQN-based MCS selection algorithm can learn perfectly almost the knowledge of the quasi-static interference.

Figure 2.5 shows the average transmission rates of different algorithms in the dynamic-interference scenario, in which the average SNR $\frac{P_p \bar{g}_p}{\sigma^2}$ received at the BS is 20 dB and average interference-to-noise ratio $\frac{P_k \bar{g}_k}{\sigma^2}$ ($k \in \{1, 2, 3\}$) received at the BS is 5 dB. From the figure, we observe that the optimal MCS selection can achieve an average transmission rate of around 2.9 kbits/frame, the UCB learning algorithm can converge to an average transmission rate of around 2 kbits/frame, and the SNR-based algorithm can achieve an average transmission rate of around 1.3 kbits/frame. Meanwhile, the proposed DQN-based MCS selection algorithm can converge to an average transmission rate of around 2.6 kbits/frame. Thus, the average transmission rate performance of the proposed DQN-based MCS selection algorithm is around 90% of the optimal MCS selection, and is around 30% higher than the UCB learning algorithm, and is 100% higher than the SNR-based algorithm. This figure demonstrates the effectiveness of the proposed DQN-based MCS selection algorithm when the interference from secondary users to the BS is highly dynamic.

2.5.3   Performance with Different Φ

Figure 2.6 shows the average transmission rate of the proposed DQN-based MCS selection algorithm with different Φ in the quasi-static interference scenario, in which the SNR at the BS, the number of secondary user pairs, interference probability of each secondary user pair, and the interference-to-noise ratio from each secondary user pair to the BS are the same as those in Fig. 2.4. From the figure, we observe that the average transmission rate of the proposed DQN-based MCS selection algorithm is almost constant as Φ grows from 1 to 10. This is reasonable. The interference pattern at the BS is dominated by the interference channel gain variation pattern. In the quasi-static interference scenario, each interference channel gain changes slowly. Thus, the data in more historical frames cannot provide more information about the interference pattern for the DQN agent to infer the upcoming interference. This result implies that it is unnecessary to consider the historical data of multiple previous frames as the environmental state when the interference at the BS is quasi-static.

Figure 2.7 shows the average transmission rate of the proposed DQN-based MCS selection algorithm with different Φ in the dynamic interference scenario, in which the SNR at the BS, the secondary user pair number, the interference probability of

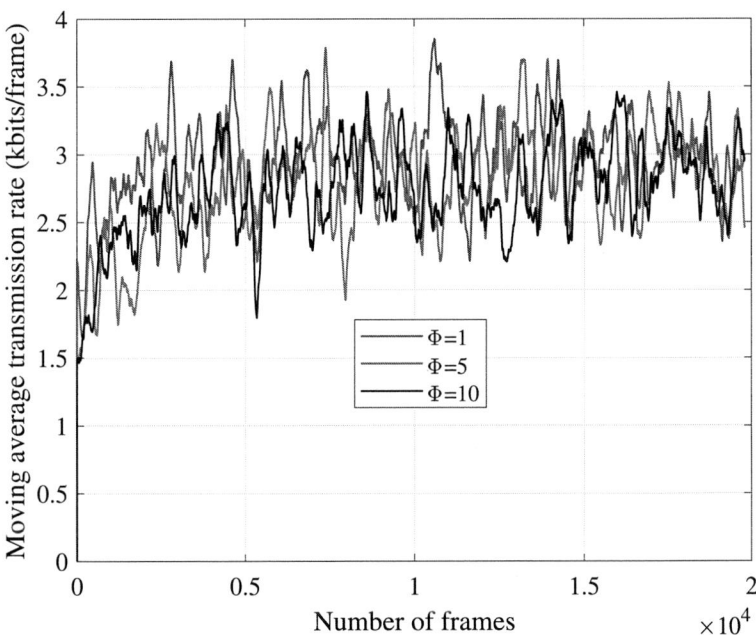

Fig. 2.6 Average transmission rate of the proposed algorithm with different Φ in the quasi-static interference scenario. Each value is a moving average in the previous 200 frames and each curve is obtained by averaging 20 independent trials

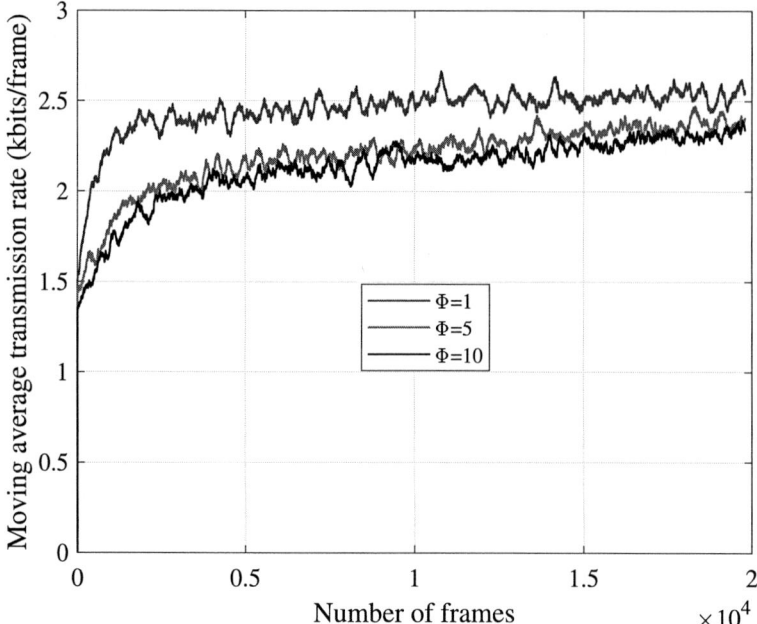

Fig. 2.7 Average transmission rate of the proposed algorithm with different Φ in the dynamic interference scenario. Each value is a moving average in the previous 200 frames and each curve is obtained by averaging 20 independent trials

each secondary user pair, and the interference-to-noise ratio from each secondary user pair to the BS are the same as those in Fig. 2.5. From the figure, we observe that the average transmission rate of the proposed DQN-based MCS selection algorithm drops as Φ grows from 1 to 10. This is reasonable. As aforementioned, the interference pattern at the BS is dominated by the interference channel gain variation pattern. The considered channel model is a first-order Markov process, meaning that each interference channel gain is only related to the interference channel gain in the previous adjacent frame. Note that in the dynamic interference scenario, each interference channel gain changes rapidly. Then, the data in previous frames cannot provide more information about the interference pattern for the DQN agent to infer the upcoming interference, but instead causes confusion to the DQN agent and degrades the performance. This result implies that it is harmful to include the historical data in multiple previous frames in each state when the interference to the BS is highly dynamic.

2.5.4   The Impact of Switching Cost on the Performance

Figure 2.8 compares the converged transmission rates and the switching rates with different switching costs of the proposed DQN-based MCS selection algorithm in a quasi-static interference scenario. For comparison, we also provide the performance of the optimal MCS selection algorithm, the UCB learning algorithm, and the SNR-based algorithm. Here, the SNR at the BS, the secondary user pair number, the interference probability of each secondary user pair, and the interference-to-noise ratio from each secondary user pair to the BS are the same as those in Fig. 2.4. Note that, the optimal algorithm and the SNR-based algorithm are independent from the switching cost and thus remain constant as the switching cost changes. Besides, the average converged transmission rate of the UCB algorithm decays from around 2.15 to 1.4 in kbits/frames as the switching cost grows from 0 to 6, and the corresponding switching rate remains around 0.03. The converged average transmission rate of the proposed DQN-based algorithm decays from around 3 to 2.4 in kbits/frames as the switching cost grows from 0 to 6, and the corresponding switching rate decays from around 0.26 to around 0.03. Thus, by adjusting the switching cost, the converged average transmission rate and switching rate of the DQN-based algorithm can simultaneously outperform the SNR-based algorithm. Besides, by adjusting the switching cost, the converged average transmission rate and switching rate of the

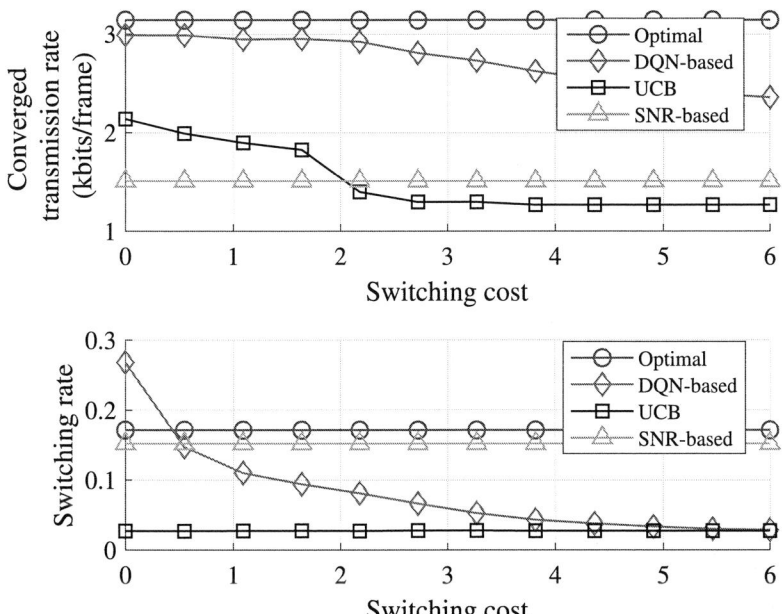

Fig. 2.8 Converged average transmission rate and switching rate with different switching costs in the quasi-static interference scenario

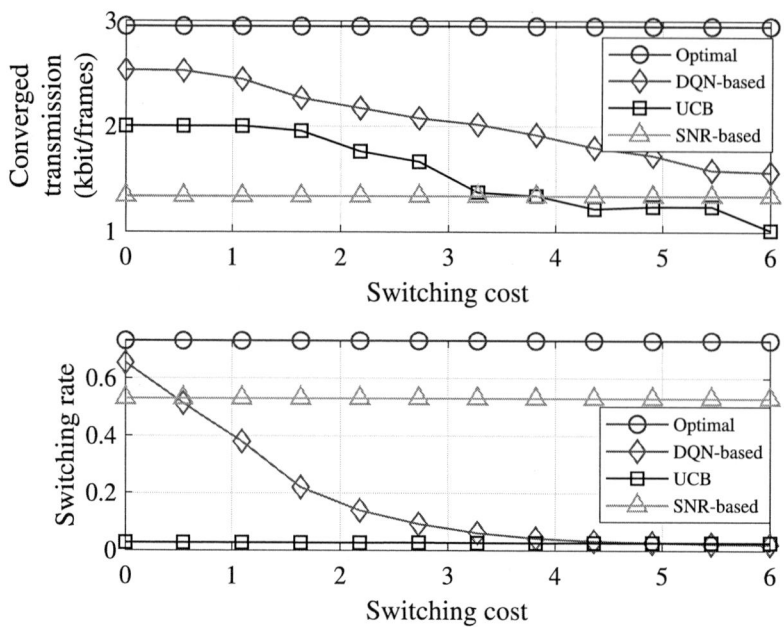

Fig. 2.9 Converged transmission rates and switching rates of different algorithms in a dynamic interference scenario

DQN-based algorithm can outperform the UCB algorithm without increasing the switching rate.

Figure 2.9 compares the converged transmission rates and the switching rates with different switching costs of the proposed DQN-based MCS selection algorithm in a quasi-static interference scenario. For comparison, we also provide the performance of the optimal MCS selection algorithm, the UCB learning algorithm, and the SNR-based algorithm. Here, the SNR at the BS, the secondary user pair number, the interference probability of each secondary user pair, and the interference-to-noise ratio from each secondary user pair to the BS are the same as those in Fig. 2.5. In general, the trend of each curve in Fig. 2.9 is similar to that in Fig. 2.8. Specifically, by adjusting the switching cost, the converged average transmission rate and switching rate of the DQN-based algorithm can simultaneously outperform the SNR-based algorithm. Besides, by adjusting the switching cost, the converged average transmission rate and switching rate of the DQN-based algorithm can outperform the UCB algorithm without increasing the switching rate.

2.6 Conclusions

In this chapter, we proposed an intelligent DQN-based MCS selection algorithm for primary transmissions in a cognitive edge network. By enabling the BS to learn proactively the interference pattern from secondary users, the BS is able to predict the upcoming interference and select a proper MCS to boost the transmission performance. To avoid significant system overheads caused by frequent MCS switchings, we introduce a switching cost, which can be adapted flexibly. Without the switching cost design, the proposed algorithm can achieve $90 \sim 100\%$ of the average transmission rate of the optimal MCS selection scheme, and has 30% average transmission rate gain compared with the UCB algorithm, and has 100%average transmission rate gain compared with the SNR-based algorithm. With the switching cost design, the proposed algorithm outperforms the benchmark algorithms in terms of the average transmission rate while maintaining comparable system overhead levels.

References

1. Wang, Y.-P.E., et al.: A primer on 3GPP narrowband Internet of Things (NB-IoT). IEEE Commun. Mag. **55**(3), 117–123 (2017)
2. Zhang, L., Xiao, M., Wu, G., Li, S.: Efficient scheduling and power allocation for D2D-assisted wireless caching networks. IEEE Trans. Commun. **64**(6), 2438–2452 (2016)
3. Zhang, L., Liang, Y.-C., Xiao, M.: Spectrum sharing for Internet of Things: a survey. IEEE Wirel. Commun. (2018). https://doi.org/10.1109/MWC.2018.1800259
4. Yang, C., Li, J., Guizani, M., Anpalagan, A., Elkashlan, M.: Advanced spectrum sharing in 5G cognitive heterogeneous networks. IEEE Wirel. Commun. vol. **23**(2), 94–101 (2016)
5. Zhang, L., Liu, J., Xiao, M., Wu, G., Liang, Y.-C., Li, S.: Performance analysis and optimization in downlink noma systems with cooperative full-duplex relaying. IEEE J. Sel. Areas Commun. **35**(10), 2398–2412 (2017)
6. Hu, R.Q., Qian, Y.: An energy efficient and spectrum efficient wireless heterogeneous network framework for 5G systems. IEEE Commun. Mag. **52**(5), 94–101 (2014)
7. Zhang, Q., Guo, H., Liang, Y.-C., Yuan, X.: Constellation learning based signal detection for ambient backscatter communication systems. IEEE J. Sel. Areas Commun. **37**(2), 452–463 (2019)
8. Zhang, Q., Zhang, L., Liang, Y.-C., Kam, P.Y:. Backscatter-NOMA: a symbiotic system of cellular and Internet-of-Things networks. IEEE Access **7**(1), 20000–20013 (2019)
9. Yang, G., Liang, Y.-C., Zhang, R., Pei, Y.: Modulation in the air: backscatter communications over ambient OFDM signals. IEEE Trans. Commun. **66**(3), 1219–1233 (2018)
10. Yang, G., Zhang, Q., Liang, Y.-C.: Cooperative ambient backscatter communication systems for future Internet-of-Things. IEEE Internet Things J. **5**(2), 1116–1130 (2018)
11. ElSawy, H., Hossain, E., Kim, D.I.: HetNets with cognitive small cells: user offloading and distributed channel allocation techniques. IEEE Commun. Mag. **51**(6), 28–36 (2013)
12. Andrews, J.G., Baccelli, F., Ganti, R.K.: A tractable approach to coverage and rate in cellular networks. IEEE Trans. Commun. **59**(11), 3122–3134 (2011)
13. Cheung, W., Quek, T.Q.S., Kountouris, M.: Throughput optimization, spectrum allocation, access control in two-tier femtocell networks. IEEE J. Sel. Areas Commun. **30**(3), 561–574 (2012)

14. Mukherjee, S.: Distribution of downlink SINR in heterogeneous cellular networks. IEEE J. Sel. Areas Commun. **30**(3), 575–585 (2012)
15. Lien, S.-Y., Chen, K.-C., Liang, Y.-C., Lin, Y.: Cognitive radio resource management for future cellular networks. IEEE Wirel. Commun. **21**(1), 70–79 (2014)
16. Liang, Y.-C., Zeng, Y., Peh, E.C.Y., Hoang, A.T.: Sensing-throughput tradeoff for cognitive radio networks. IEEE Trans. Wirel. Commun. **7**(4), 1326–1337 (2008)
17. Zhang, L., Xiao, M., Wu, G., Alam, M., Liang, Y.-C., Li, S.: A survey of advanced techniques for spectrum sharing in 5G networks. IEEE Wirel. Commun. **24**(5), 44–51 (2017)
18. Luong, N.C., Hoang, D.T., Gong, S., Niyato, D., Wang, P., Liang, Y.-C., Kim, D.I.: Applications of deep reinforcement learning in communications and networking: a survey. Available in arXiv:1810.07862 [cs.NI]
19. Zhang, L., Xiao, M., Wu, G., Zhao, G., Liang, Y.C., Li, S.: Energy-efficient cognitive transmission with imperfect spectrum sensing. IEEE J. Sel. Areas Commun. **34**(5), 1320–1335 (2016)
20. Kim, T., Love, D.J., Clerckx, B.: Does frequent low resolution feedback outperform infrequent high resolution feedback for multiple antenna beamforming systems? IEEE Trans. Signal Process. **59**(4), 1654–1669 (2011)
21. Alnwaimi, G.R., Hatem, B.: Adaptive packet length and MCS using average or instantaneous SNR. IEEE Trans. Veh. Technol. (2018) (early access)
22. Mnih, V., et al.: Human-level control through deep reinforcement learning. Nature **518**, 529–533 (2015)
23. Farrokh, A., Krishnamurthy, V., Schober, R.: Optimal adaptive modulation and coding with switching costs. IEEE Trans. Commun. **57**(3), 697–706 (2009)
24. Wang, Z., Li, L., Xu, Y., Tian, H., Cui, S.: Handover control in wireless systems via asynchronous multi-User deep reinforcement learning. IEEE Internet Things J. (2018) (early access)
25. Shen, C., Teki, C., van der Schaar, M.: A non-stochastic learning approach to energy efficient mobility management. IEEE J. Sel. Areas Commun. **34**(12), 3854–3868 (2016)
26. Proakis, J.G.: Communications. McGraw-Hill, New York (2008)
27. Nasir, Y.S., Guo, D: Deep reinforcement learning for distribute dynamic power allocation in wireless networks. Available in arXiv:1808.00490 [eess.SP]
28. Tieleman, T., Hinton, G.: Lecture 6.5-rmsprop: divide the gradient by a running average of its recent magnitude. COURSERA: Neural Netw. Mach. Learn. **4**(2), 26–31 (2012)

Chapter 3
AI-Enabled Centralized Spectrum Sharing

Abstract This chapter considers the AI-enabled centralized spectrum sharing in the wireless edge network, where multiple access points (APs) are deployed to serve users by reusing the same spectrum band. To alleviate the severe interference caused by spectrum reusing, advanced power control techniques are introduced to manage the interference and improve the sum-rate of the whole network. Conventional power control techniques require instantaneous global channel state information (CSI) to optimize the power control strategy. Nevertheless, instantaneous global CSI is impractical to acquire due to the fast-changing nature of the time-variable channel. To address this problem, this chapter leverages DRL to design a centralized learning algorithm catering to the power control. Specifically, by establishing a local DNN at each AP, we propose a multiple-actor-shared-critic (MASC) method to train all local DNNs in a centralized manner in the cloud. Then, each AP can determine the transmit power with the well-trained local DNN together with the local observations. Simulation results demonstrate that the performance of the proposed algorithm exceeds the benchmarks in terms of sum-rate and time complexity.

Keywords DRL · Multi-agent · Power control · MASC · Het-Net

3.1 Introduction

Wireless data traffics have increased rapidly as the number of ubiquitous wireless devices like smartphones and tablets exponentially grows [1–3]. To offload the heavy wireless traffics of the macro BS for providing wireless access services for all the users within the macro-cell, *heterogeneous network* (HetNet) is introduced as a main component of the wireless edge network by designing small cells in the macro-cell. Unfortunately, it is impractical to assign orthogonal spectrum resources to all the cells (including macro-cell and small cells) due to the spectrum scarcity issue, and the same spectrum reusing scenario among different cells inevitably brings severe inter-cell interference. Therefore, power control techniques have been proposed to alleviate the severe inter-cell interference and enhance the sum-rate of the cells [4–12].

Provided that the instantaneous global (including both intra-cell and inter-cell) *channel state information* (CSI) is available, conventional power control algorithms, such as *weighted minimum mean square error* (WMMSE) algorithm [6] and *fractional programming* (FP) [7] algorithm, optimize the power control policies for the *access points* (APs) in different cells simultaneously. However, it is challenging to collect the instantaneous global CSI in HetNet, and the optimized policies of the conventional algorithms may be sub-optimal or outdated resulting from the ultra-dynamic characteristic of the wireless channels together with the long processing time of the conventional algorithms. These limitations become the bottleneck to enhance the sum-rate of the whole network.

Recently, RL [13–20] and DRL [21–33] have been introduced in wireless network design because of their appealing performance in the computer science field. Thanks to the strong representation capability of the DRL, there is some literature utilizing the DRL to learn the optimal power control policy with only historical data [11, 12]. Compared to the instantaneous global CSI, historical global CSI is easy to collect by gathering the local CSI from all APs in the core network. In addition, the historical global CSI inherits useful long-term information that can be directly utilized to optimize the power control policy. Inspired by these, in this chapter, we adopt the DRL to design a multi-agent power control algorithm with centralized learning in a HetNet. With the proposed algorithm, the optimal power control policy can be learned with the historical global CSI in a centralized manner, and each AP can determine the optimal transmit power utilizing only the instantaneous local CSI with low time complexity.

3.2   Syetem Model

Figure 3.1 demonstrates the considered spectrum sharing scenario in a wireless edge network, in which N APs share the same spectrum band to serve users and may cause interference with each other. If we denote an AP as AP n, where $n \in \mathbb{N} = \{1, 2, \cdots, N\}$. Then, the user served by AP n can be denoted as *user equipment* (UE) n. The channel model and signal transmission mode are illustrated as follows.

3.2.1   Channel Model

We consider the large-scale attenuation (including path-loss and shadowing) and small-scale block Rayleigh fading to simulate the realistic channel between the AP and UE. The channel gain can be represented by $g_{n,k} = \phi_{n,k}|h_{n,k}|^2$, where $\phi_{n,k}$ and $h_{n,k}$ denote the large-scale attenuation and the small-scale block Rayleigh fading from AP n to UE k, respectively. It is worth noting that the large-scale attenuation is highly related to the distance between the AP and the UE. The small-

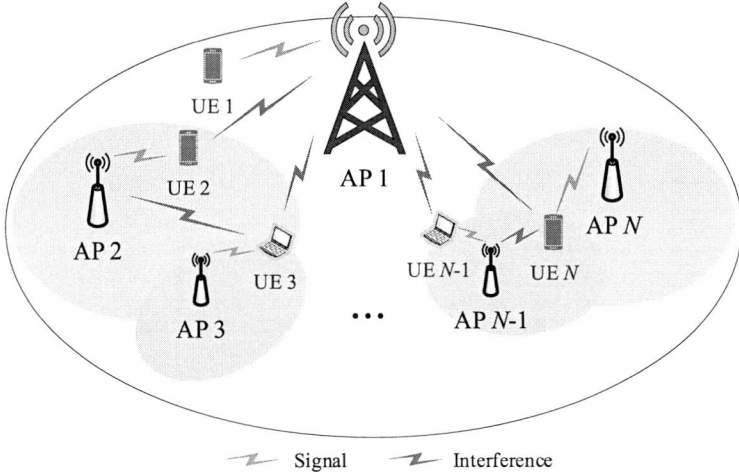

Fig. 3.1 Typical spectrum sharing scenario in a HetNet, which consists of multiple APs sharing the same spectrum band to the served users within their coverages and may cause interference to each other

scale block Rayleigh fading remains constant in each single time slot and changes among different time slots.

Jake's model is adopted to represent the relationship between the small-scale Rayleigh fadings in two successive time slots [34]. The small-scale block Rayleigh fading can be represented by

$$h(t) = \rho h(t-1) + \omega, \tag{3.1}$$

where ρ $(0 \leq \rho \leq 1)$ denotes the correlation coefficient of small-scale Rayleigh fading realizations between two successive time slots, ω and $h(0)$ are in random distributions, i.e., $\omega \sim \mathcal{CN}(0, 1 - \rho^2)$ and $h(0) \sim \mathcal{CN}(0, 1)$, respectively. In particular, Jake's model reduces to the *independent and identically distributed* (IID) channel model if ρ is zero.

3.2.2   Signal Transmission Model

According to the previous channel model, the received signal at UE n at time slot t can be represented by

$$
\begin{aligned}
y_n(t) = & \sqrt{p_n(t)\phi_{n,n}}h_{n,n}(t)x_n(t) \\
& + \sum_{k \in \mathbb{N}, k \neq n} \sqrt{p_k(t)\phi_{k,n}}h_{k,n}(t)x_k(t) + \delta_n(t),
\end{aligned} \tag{3.2}
$$

where $p_n(t)$ denotes the transmit power from AP n to UE n at time slot t, $x_n(t)$ denotes the downlink signal from AP n to UE n with unit power at time slot t, $\delta_n(t)$ is the noise at UE n with power σ^2. Especially, the first term represents the received signal from AP n, and the second term implies the interference from other APs caused by frequency sharing. Therefore, the SINR at UE n can be calculated by

$$\gamma_n(t) = \frac{p_n(t)g_{n,n}(t)}{\sum_{k \in \mathbb{N}, k \neq n} p_k(t)g_{k,n}(t) + \sigma^2}. \tag{3.3}$$

Then, if we denote B as the bandwidth, the downlink transmission rate (in bps) from AP n to UE n is

$$r_n(t) = B \log\left(1 + \gamma_n(t)\right). \tag{3.4}$$

3.3   Problem Description and Analysis

Since the frequency-sharing scenario may cause severe interference, our goal is to optimize the transmit power of all APs to alleviate the interference, thus, maximizing the overall downlink sum-rate. The optimization problem can be formulated as

$$\max_{p_n, \forall n \in \mathbb{N}} R(t) = \sum_{n=1}^{N} r_n(t)$$

$$\text{s.t. } 0 \leq p_n(t) \leq p_{n,\max}, \forall n \in \mathbb{N}, \tag{3.5}$$

where $p_{n,\max}$ denotes the maximum transmit power constraint of AP n. In particular, different types of APs have different maximum transmit power constraints.

The considered optimization problem brings about two main challenges: (1) The considered problem is known to be NP-hard according to [35]. This implies the optimal transmit power is hard to determine. (2) The optimal transmit power that maximizes the overall sum-rate should be determined at the beginning of each time slot with low time complexity. Due to the dynamic characteristic of wireless channels, the determined transmit power with high complexity may become stale and degrade the overall sum-rate greatly.

According to the previous discussion, conventional power control algorithms, such as the WMMSE and FP algorithms, generally assume that the communication environment is quasi-static, where wireless channels change slowly. However, these algorithms are not applicable in a dynamic communication environment. Consequently, DRL [12] is introduced to address this problem by assuming the acquisition of the instantaneous global CSI is available and the cooperations among neighboring APs are guaranteed. Indeed, all of the mentioned algorithms are inapplicable in the considered scenario of this paper due to the following two main constraints:

- Instantaneous global CSI is unavailable in HetNet.
- Neighboring APs are not willing to or even cannot cooperate.

Therefore, we intend to develop a DRL based multi-agent power control algorithm to solve the above problem.

3.4   DRL Based Multi-Agent Power Control Algorithm

It is known that the optimal transmit power of APs depends on the instantaneous global CSI. Nevertheless, the acquisition of the instantaneous global CSI is impractical in HetNet. We notice that the historical wireless data of the whole network (e.g., global CSI, transmit power of APs, mutual interference, and achieved sum-rates) can be easily collected and gathered. These data contain useful long-term information that can be directly utilized to optimize the transmit power of APs instead of the instantaneous global CSI. Consequently, by leveraging the DRL, we develop an intelligent power control algorithm that can fully utilize the historical wireless data of the whole network. In this section, we will first outline the framework of the proposed intelligent power control algorithm, introduce its detailed design, and describe its implementation.

3.4.1   Framework Outline

We adopt a centralized-training-distributed-execution architecture as the basic algorithm framework as shown in Fig. 3.2. In general, the framework consists of N local DNNs deployed APs to determine the optimal transmit power, and a core network to train all of the local DNNs. Specifically, a local DNN is established at each AP, where the input and output of each local DNN are instantaneous local information and optimal transmit power, respectively. The core network collects historical wireless data of the whole network, and utilizes the historical data to train the weight parameters of all local DNNs. In this way, each local DNN can acquire the long-term knowledge of the whole network in terms of power control, and make instant distributed decisions with only real-time local information.

Additionally, we develop a novel MASC training method based on the DDPG to update the weight vectors of local DNNs from the DNNs in the core network, as shown in Fig. 3.2. Specifically, N actor DNNs together with N target actor DNNs are established in the core network corresponding to N local DNNs. It's worth noting that each actor/target actor DNN has an identical structure as the associated local DNN. This enables each local DNN to directly update its parameters with trained ones of the actor DNN. Meanwhile, a shared critic DNN and a target critic

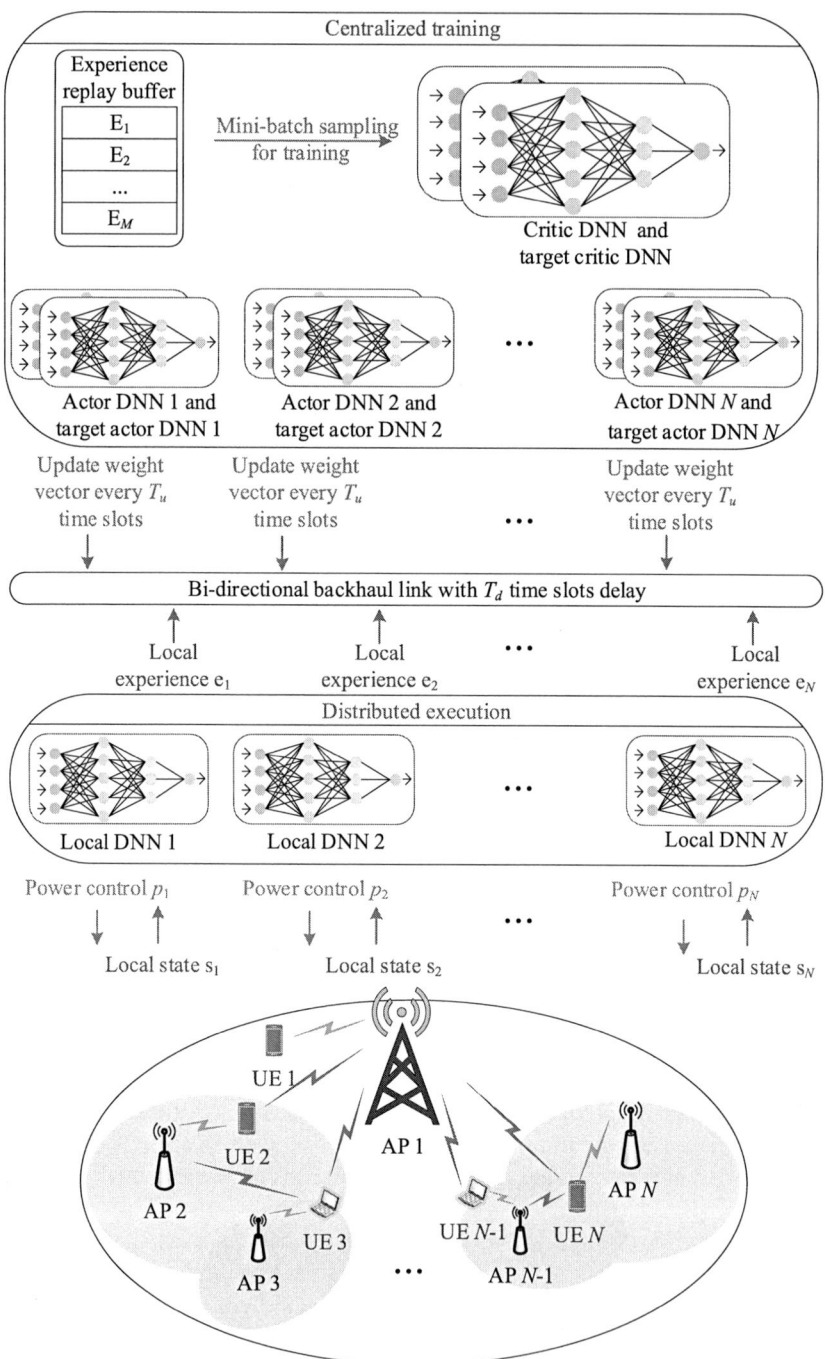

Fig. 3.2 Proposed algorithm framework

DNN with the same structure are also established in the core network to train all the actor DNNs. The input of the critic DNN is the historical local information of all APs mixed up with their adopted transmit power, and the output is the long-term sum-rate that evaluates the performance of all the actor DNNs. In this way, by constantly training each actor DNN with the evaluation of the critic DNN, the weight parameters of each actor DNN can be updated toward the direction of the global optimum. Eventually, the trained parameters of all actor DNNs can be utilized to update the corresponding local DNNs until convergence.

The proposed framework distinguishes itself in solving the power control problem in HetNet, and can be summarized into two key advantages:

- Each AP can utilize the well-trained local DNN to determine the optimal transmit power at the beginning of each time slot with low time complexity, guaranteeing the timeliness of the design.
- Each AP can independently optimize the transmit power in terms of the enhanced sum-rate with only local information, avoiding the acquisition of global CSI.

In the following sections, we introduce the algorithm design regarding the design of local DNNs, (target) actor DNNs, and critic DNN, respectively.

3.4.2 Algorithm Design

3.4.2.1 Design of the Local/Actor DNN

The design of the local DNN can be divided into action space, state space, local experience and structure.

Action Space The action space is designed as the transmit power of AP n. If we denote $a_n(t)$ as the action space of local DNN n, then we have $a_n(t) = p_n(t)$.

State Space If we denote s_n as the state space of local DNN n, the state space can be represented by

$$
s_n(t) = \left\{ g_{n,n}(t-1), p_n(t-1), \sum_{k \in \mathbb{N}, k \neq n} p_k(t-1)g_{k,n}(t-1), \right.
$$
$$
\left. \gamma_n(t-1), r_n(t-1), g_{n,n}(t), \sum_{k \in \mathbb{N}, k \neq n} p_k(t-1)g_{k,n}(t) \right\}, \tag{3.6}
$$

where $g_{n,n}(t-1)$ is the historical channel gain between the AP n and UE n at time slot $t-1$, $p_n(t-1)$ is the historical transmit power of AP n at time slot $t-1$, $\sum_{k \in \mathbb{N}, k \neq n} p_k(t-1)g_{k,n}(t-1)$ is the historical sum-interference from AP k ($k \in \mathbb{N}, k \neq n$) at time slot $t-1$, $\gamma_n(t-1)$ is the received SNR at UE n at time slot $t-1$, $r_n(t-1)$

is the historical transmission rate at UE n at time slot $t - 1$. Similarly, $g_{n,n}(t)$ is the current channel gain between AP n and UE n at time slot t, $\sum_{k \in \mathbb{N}, k \neq n} p_k(t-1) g_{k,n}(t)$ is the current sum-interference from AP k. It is worth noting that, the historical state information can be directly acquired at the current time slot, while the current state information at UE n is generated with the historical transmit power of AP n, i.e., $p_n(t-1)$, even though CSI of the whole HetNet has changed at the beginning of the current time slot. This is because the new transmit power can not be determined and configured as soon as the CSI changes.

Reward If we denote $r_n(t)$ as the reward of the AP n, it equals the achievable rate of AP n and can be calculated by (3.4).

Local Experience The experience $e_n(t)$ can be constructed as

$$e_n(t) = \{s_n(t - 1), a_n(t - 1), r_n(t - 1), s_n(t)\}. \tag{3.7}$$

Network Structure As shown in Fig. 3.3a, the designed structure of each local/actor DNN consists of five full-connected layers. Specifically, the first layer is the input layer that the $L_1^{(a)} = 7$ neurons corresponding to the input s_n with 7 elements. The second layer and the third layer are hidden layers that have $L_2^{(a)}$ and $L_3^{(a)}$ neurons, respectively. The fourth layer is the normalized layer with the sigmoid activation function including $L_4^{(a)} = 1$ neuron. This layer is designed to normalize the output value between zero and one. Eventually, the fifth layer is the output layer with $L_5^{(a)} = 1$ neuron, the function of which is to scale the output value from the fourth layer to a value between zero and $p_{n,max}$. To this end, the designed structure of each local/actor DNN takes the local state s_n as the input, and guarantees the output is a transmit power that satisfies the maximum transmit power constraint.

3.4.2.2   Design of the Critic DNN

The critic DNN is designed to train all the actor DNNs with historical information. Specifically, the input is the global state mixed up with the global action of all APs suffering from the transmission delay between the APs and the cloud. The output is the whole network's expected long-term sum-rate, which evaluates the action performances of all APs. The design of critic DNN can be divided into global action, global state, global reward, global experience and structure.

Global Action Space The global action space consists of the action spaces of all APs with the transmission delay T_d. If we denote $A(t)$ as the global action space, it can be represented by $A(t) = \{a_1(t - T_d), \cdots, a_N(t - T_d)\}$.

Global State Space The global state space consists of state spaces of all APs together with other global information, i.e., coordinates of APs and UEs. Here, we denote other global information as $s_o(t)$, and denote $S(t)$ as the global state space.

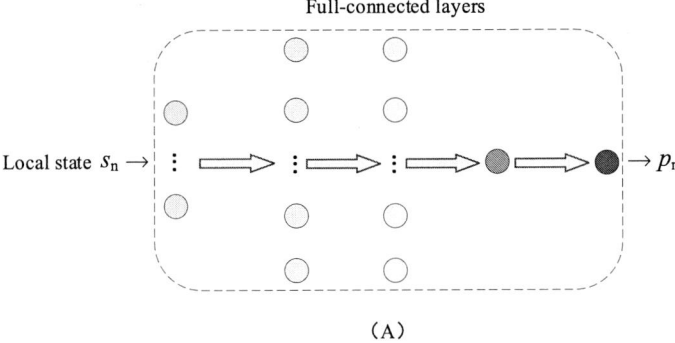

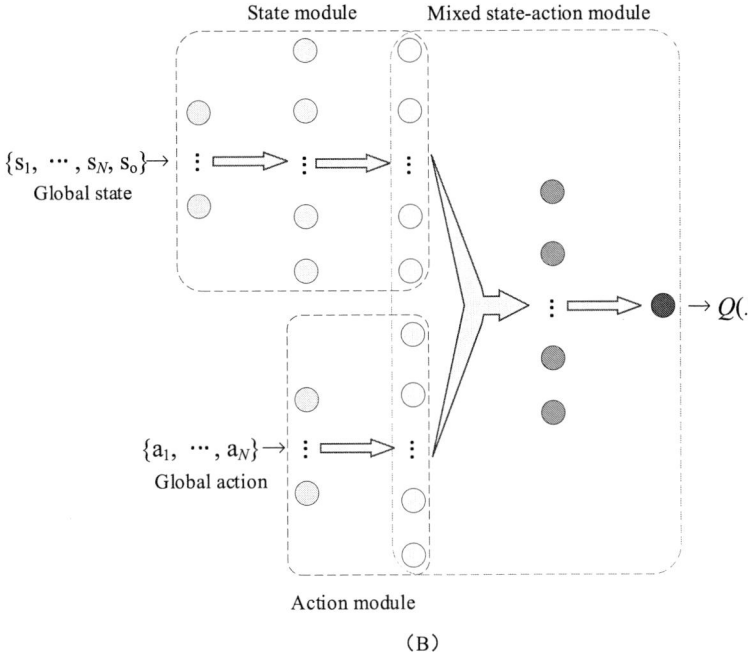

Fig. 3.3 The structures of each actor DNN and the critic DNN, in which the arrows indicate the direction of the data flow. Each dotted box contains a certain number of hidden layers. (**a**) Structure of each local/actor DNN. (**b**) Structure of the critic DNN

Then, the global state space can be represented by $S(t) = \{s_1(t - T_d), \cdots, s_N(t - T_d), s_0(t - T_d)\}$.

Global Reward If we denote $R_n(t)$ as the reward, it is designed as the summation of achievable rate for all APs and can be calculated by, i.e., $R_n(t - T_d) = \sum_{n \in \mathbb{N}} r_n(t - T_d)$.

Global Experience If we denote $E(t)$ as the global experience, it can be represented by

$$E(t) = \{s_1(t - 1 - T_d), \cdots, s_N(t - 1 - T_d), s_0(t - 1 - T_d),$$
$$a_1(t - 1 - T_d), \cdots, a_N(t - 1 - T_d), R(t - 1 - T_d),$$
$$s_1(t - T_d), \cdots, s_N(t - T_d), s_0(t - T_d)\}. \tag{3.8}$$

It is worth noting that $s_n(t - 1 - T_d)$, $s_n(t - T_d)$, $a_n(t - 1 - T_d)$, $a_n(t - 1 - T_d)$ ($\forall n \in \mathbb{N}$) and $R(t - 1 - T_d)$ can be directly acquired or calculate from the local experience of all APs.

Network Structure As illustrated in Fig. 3.3b, there are three modules in the designed critic DNN, i.e., state module, action module, and mixed state-action module. Each module is made up of several cascaded fully-connected layers. To begin with, the state module consists of three full-connected layers. The first layer is the input layer for the global state $\{s_1, \cdots, s_N, s_0\}$. It is worth noting that there are seven elements in each s_n ($n \in \mathbb{N}$), and the whole network includes N s_n corresponding to N APs. In addition, there are N^2 elements in s_0. Therefore, the input layer is with $L_1^{(S)} = 7N + N^2$ neurons in total. Then, The second and the third layer are with $L_2^{(S)}$ and $L_3^{(S)}$ neurons, respectively. Then, the action module consists of two full-connected layers, where the first layer is the input layer with $L_1^{(A)} = N$ neurons, corresponding to the input a_n ($n \in \mathbb{N}$). The second layer of the action module is with $L_2^{(A)}$ neurons. As for the mixed state-action module, it has three full-connected layers. The first layer is the input layer, which concatenates the last layers of the state module and the action module. Therefore, there are $L_1^{(M)} = L_3^{(S)} + L_2^{(A)}$ neurons in the first layer. The second layer of the mixed state-action module is a hidden layer and has $L_2^{(M)}$ neurons. The third layer is the output layer with one neuron, which outputs the long-term reward $Q^{(c)}(s_1, \cdots, s_N, s_0, a_1, \cdots, a_N; \theta^{(c)})$. In summary, there are $N^2 + 8N + 1 + L_2^{(S)} + L_3^{(S)} + L_2^{(A)} + L_2^{(M)}$ neurons in the critic DNN.

3.4.2.3   Algorithm Implementation

For simplicity, we denote the N local DNNs, actor DNNs, and target actor DNNs as $\mu_n^{(L)}\left(s_n; \theta_n^{(L)}\right)$, $\mu_n^{(a\text{-})}\left(s_n; \theta_n^{(a\text{-})}\right)$, and $\mu_n^{(a)}\left(s_n; \theta_n^{(a)}\right)$, ($n \in \mathbb{N}$), respectively, where s_n is the local state of AP n, $\theta_n^{(L)}$, $\theta_n^{(a)}$ and $\theta_n^{(a\text{-})}$ denotes the weight parameters of the local DNN, actor DNN, target actor DNN n, respectively. Moreover, $Q\left(s_1, \cdots, s_N, s_0, a_1, \cdots, a_N; \theta^{(c)}\right)$ and $Q^-\left(s_1, \cdots, s_N, s_0, a_1, \cdots, a_N; \theta^{(c\text{-})}\right)$ denotes the critic DNN and the target critic DNN, respectively, where s_n ($n \in \mathbb{N}$) denotes the local states of AP n and s_0 denotes another global state of the whole

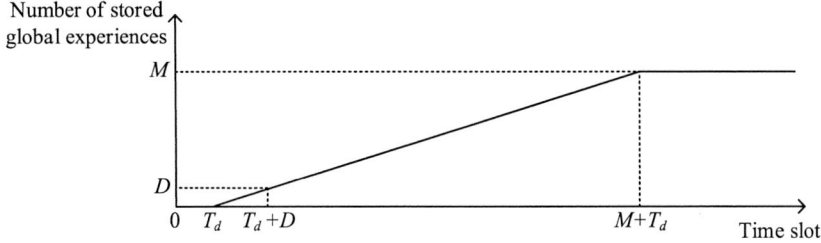

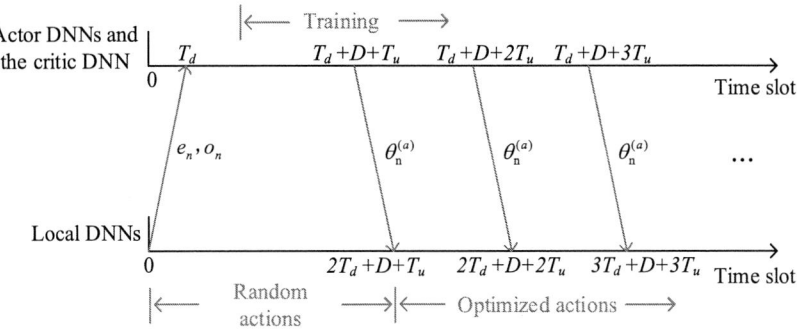

Fig. 3.4 Diagram of the proposed algorithm

network, a_n ($\forall n \in \mathbb{N}$) denotes action space of AP n, $\theta^{(c)}$ and $\theta^{(c\text{-})}$ are the weight parameters of critic and target critc DNN, respectively.

The proposed algorithm can be divided into three stages, i.e., initialization stage, random experience accumulation stage, and repeat stage. The proposed algorithm is summarized in Algorithm 1, and the diagram of the proposed algorithm is summarized in Fig. 3.4.

In the initialization stage, N local DNNs, N actor DNNs, N target actor DNNs, a critic DNN, and a target critic DNN are first established and initialized. Specifically, local DNN $\mu_n^{(L)}(s_n; \theta_n^{(L)})$ ($\forall n \in \mathbb{N}$) is established at AP n with the structure in Fig. 3.3a. The actor DNN $\mu_n^{(a)}(s_n; \theta_n^{(a)})$ and the corresponding target actor DNN $\mu_n^{(a\text{-})}(s_n; \theta_n^{(a\text{-})})$ are established in the core network with the identical structure as the local DNN n. Meanwhile, a critic DNN $Q^{(c)}(s_1, \cdots, s_N, s_0, a_1, \cdots, a_N; \theta^{(c)})$ and the corresponding target critic DNN $Q^{c\text{-}}(s_1, \cdots, s_N, s_0, a_1, \cdots, a_N; \theta^{(c\text{-})})$ are established in the core network with the structure in Fig. 3.3b. Then, $\theta_n^{(L)}$, $\theta_n^{(a)}$, and $\theta^{(c)}$ are randomly initialized, and $\theta_n^{(a\text{-})}$ and $\theta^{(c\text{-})}$ are initialized with $\theta_n^{(a)}$ and $\theta^{(c)}$, respectively.

In the random experience accumulation stage ($t \leq 2T_d + D + T_u$), at the beginning of time slot t, UE n ($\forall n \in \mathbb{N}$) first observes local state $s_n(t)$ and auxiliary information $o_n(t)$, and transmits them to AP n. After that, AP randomly determines an action (i.e., transmit power) and uploads the local experience $e_n(t)$ and $o_n(t)$ to

---

Algorithm 2 Proposed DRL based multi-agent power control algorithm.

1: **Initialization:**
2: Establish required DNN networks, i.e., $\mu_n^{(L)}(s_n; \theta_n^{(L)})$, $\mu_n^{(a)}(s_n; \theta_n^{(a)})$, $\mu_n^{(a-)}(s_n; \theta_n^{(a-)})$, $\mu_n^{(L)}(s_n; \theta_n^{(L)})$, $Q^{(c)}(s_1, \cdots, s_N, s_0, a_1, \cdots, a_N; \theta^{(c)})$ and $Q^{c-}(s_1, \cdots, s_N, s_0, a_1, \cdots, a_N; \theta^{(c-)})$, according to the structures in Fig. 3.3a and b.
3: Initialize $\theta_n^{(L)}$, $\theta_n^{(a)}$, and $\theta^{(c)}$ randomly, initialize $\theta_n^{(a-)}$ and $\theta^{(c-)}$ with $\theta_n^{(a)}$ and $\theta^{(c)}$, respectively.
4: **Random experience accumulation:**
5: At the beginning of time slot t $(t \leq T_d + D)$, UE n uploads $s_n(t)$ and $o_n(t)$ to AP n, and AP n randomly chooses a transmit power.
6: In time slot t $(t \leq T_d + D)$, AP n uploads local experience $e_n(t)$ and auxiliary information $o_n(t)$ to the cloud. Upon obtaining N local experiences and the corresponding N auxiliary information, the cloud can construct a global experience, which is stored in the replay buffer.
7: In time slot $t = T_d + D$, the cloud samples a mini-batch of experiences to train or update $\theta^{(c)}$, $\theta^{(c-)}$, $\theta_n^{(a)}$ and $\theta_n^{(a-)}$.
8: From time slot $t = T_d + D$, in every T_u time slots, the cloud update the weight vector $\theta_n^{(L)}$ with the latest $\theta_n^{(a)}$.
9: **Repeat:**
10: At the beginning of time slot t $(t > 2T_d + D + T_u)$, AP n sets the transmit power to be $p_n(t) = \mu_n^{(L)}(s_n; \theta_n^{(L)}) + \zeta$, where ζ is the action noise.
11: In time slot t $(t > 2T_d + D + T_u)$, the cloud samples a mini-batch of experiences to train or update $\theta^{(c)}$, $\theta^{(c-)}$, $\theta_n^{(a)}$ $(\forall n \in \mathbb{N})$, $\theta_n^{(a-)}$ $(\forall n \in \mathbb{N})$.
12: In every T_u time slots, the cloud uses the latest $\theta_n^{(a)}$ to update the weight vector $\theta_n^{(L)}$.

---

the core network through the bi-directional backhaul link with T_d time slots delay. Then, upon receiving $e_n(t)$ and $o_n(t)$ from all APs, the core network uses them to construct the global experience $E(t)$ and stores $E(t)$ into the experience replay buffer with the capacity M in a FIFO manner. As shown in Fig. 3.4, the core network accumulates D global experiences at the beginning of time slot $T_d + D$. Then, the core network starts to train the critic DNN, the target critic DNN, actor DNNs, and target actor DNNs by sampling a mini-batch of experiences $\mathcal{E}$ with length D from the memory replay buffer, and minimize $\theta^{(c)}$ of the sampled experiences, i.e.,

$$\mathbb{L}(\theta^{(c)}) = \frac{1}{D} \sum_{\mathcal{E}} [y^{\text{Tar}} - Q(s_1, \cdots, s_N, s_0, a_1, \cdots, a_N; \theta^{(c)})]^2, \qquad (3.9)$$

where y^{Tar} can be calculated by

$$y^{\text{Tar}} = \sum_{n=1}^{N} r_n + \eta \max_{a_n' \in A} Q^-(s_1', \cdots, s_N', s_0, a_1', \cdots, a_N'; \theta^{(c-)}). \qquad (3.10)$$

Training the weight parameters $\theta_n^{(a)}$ of actor DNN n is to maximize the expected long-term global reward $J(\theta_1^{(a)}, \cdots, \theta_N^{(a)})$, which is calculated by

$$
\nabla_{\theta_n^{(a)}} J(\theta_1^{(a)}, \cdots, \theta_N^{(a)})
$$

$$
\approx \frac{1}{D} \sum_{\mathcal{E}} \nabla_{a_n} Q\left(s_1, \cdots, s_N, s_o, a_1, \cdots, a_N; \theta^{(c)}\right)\Big|_{a_n = \mu(s_n; \theta_n^{(a)})}
$$

$$
\times \nabla_{\theta_n^{(a)}} \mu_n^{(a)}(s_n; \theta_n^{(a)}),
\tag{3.11}
$$

where $J(\theta_1^{(a)}, \cdots, \theta_N^{(a)})$ is defined as the expectation of $Q(s_1, \cdots, s_N, s_o, \mu_1^{(a)}$ $(s_1; \theta_1^{(a)}), \cdots, \mu_N^{(a)}(s_N; \theta_N^{(a)}); \theta^{(c)})$, and can be represented by

$$
J(\theta_1^{(a)}, \cdots, \theta_N^{(a)})
$$

$$
= \mathbb{E}_{s_1, \cdots, s_N, s_o}\left[Q\left(s_1, \cdots, s_N, s_o, \mu_1^{(a)}(s_1; \theta_1^{(a)}), \cdots, \right.\right.
$$

$$
\left.\left. \mu_N^{(a)}(s_N; \theta_N^{(a)}); \theta^{(c)}\right)\right].
\tag{3.12}
$$

As for the weight parameters of target critic DNN, denoted by $\theta^{(c-)}$, and parameters of target actor DNN n, denoted by $\theta_n^{(a-)}$, they are updated according to the soft update method adopted in DDPG, i.e,

$$
\theta^{(c-)} \leftarrow \tau^{(c)}\theta^{(c)} + (1 - \tau^{(c)})\theta^{(c-)},
\tag{3.13}
$$

$$
\theta_n^{(a-)} \leftarrow \tau_n^{(a)}\theta_n^{(a)} + (1 - \tau_n^{(a)})\theta_n^{(a-)},
\tag{3.14}
$$

where $\tau^{(c)} \in [0, 1]$ and $\tau_n^{(a)} \in [0, 1]$ denotes the learning rate of the target critic DNN and target actor DNN, respectively.

From time slot $T_d + D$, the core network transmits the latest weight parameters $\theta_n^{(a)}$ to AP n through the bi-directional backhaul link with T_d time slots delay every T_u time slots. The latest weight parameters $\theta_n^{(a)}$ arrive in AP n at time slot $2T_d + D + T_u$, then, AP n utilizes $\theta_n^{(a)}$ to update $\theta_n^{(L)}$.

In the repeat stage ($t > 2T_d + D + T_u$), UE n ($\forall n \in \mathbb{N}$) still observes $s_n(t)$ and $o_n(t)$, and transmits them to AP n. AP n utilizes the local DNN to determine and configure the transmit power, i.e., $p_n(t) = \mu_n^{(L)}(s_n(t); \theta_n^{(L)}) + \zeta$, where ζ is the random noise to guarantee a continuous exploration of the action. After that, AP n uploads local experience $e_n(t)$ and $o_n(t)$ to the core network through the bi-directional backhaul link with T_d time slots delay. Then, the core network repeats the same procedures as the random experience accumulation stage does.

3.5   Simulation Results

3.5.1   Simulation Settings

We consider both two-layer HetNet scenario and three-layer HetNet scenario:

- Two-layer HetNet scenario: As shown in Fig. 3.5a, five APs are located at $(0, 0)$, $(500, 0)$, $(0, 500)$, $(-500, 0)$, and $(0, -500)$ in meters, respectively. At the same time, the UE is randomly distributed within the coverage of each AP. The coverage of each AP is a disc area with the radius from v_{min} to v_{max}. We set $v_{min} = 10$ for all APs. For AP 1 in the first layer, we set $v_{max} = 1000$ and maximum transmit power to be 30 dBm. For AP n ($n \in \{2, 3, 4, 5\}$) in the second layer, we set $v_{max} = 200$ and the maximum transmit power to be 23 dBm.
- Three-layer HetNet scenario: As shown in Fig. 3.5b, nine APs are located at $(0, 0)$, $(500, 0)$, $(0, 500)$, $(-500, 0)$, $(0, -500)$, $(700, 0)$, $(0, 700)$, $(-700, 0)$, and $(0, -700)$ in meters, respectively. At the same time, the UE is randomly distributed within the coverage of each AP. v_{min} is set to be 10 meters for all APs. For AP 1 in the first layer, we set $v_{max} = 1000$, and set the maximum transmit power to be 30 dBm. For AP n ($n \in \{2, 3, 4, 5\}$) in the second layer, we set $v_{max} = 200$, and set the maximum transmit power to be 23 dBm. For AP n ($n \in \{6, 7, 8, 9\}$) in the third layer, we set $v_{max} = 100$, and set the maximum transmit power to be 20 dBm.

In addition, the pass-loss between the AP and UE is $120.9 + 37.6 \log 10(d)$ in dB, where d in kilometer is the distance between AP and UE [36], the log-normal shadowing standard deviation is 8 dB, the noise power σ^2 at each UE is -114 dBm, the spectrum bandwidth B is set to be 10 MHz, the capacity of the memory replay buffer M is set to be 1000, the delay T_d between the core network and the AP is set to be 50 time slots, the update period T_u is set to be $T_u = 100$ time slots. As for the hyperparameters of the DNNs, they are determined by cross-validation [12], and are summarized in Tables 3.1, 3.2 and 3.3.

3.5.2   Performance Comparison and Analysis

In this part, we evaluate the performance of the proposed algorithm in two scenarios. The proposed algorithm can be divided into two stages, i.e., training stage and testing stage. In the training stage, the weight parameters of the local DNN, actor DNN, and critic DNN are trained and updated in the first 5000 time slot. In the testing stage, the well-trained local DNN in each AP is utilized to determine the transmit power in the following 2000 time slots. It is worth noting that each curve is the average of ten independent trials. We select four benchmarks, i.e., WMMSE algorithm, FP algorithm, Full power algorithm, and Random power algorithm. Specifically, the WMMSE algorithm and the FP algorithm are initialized with the

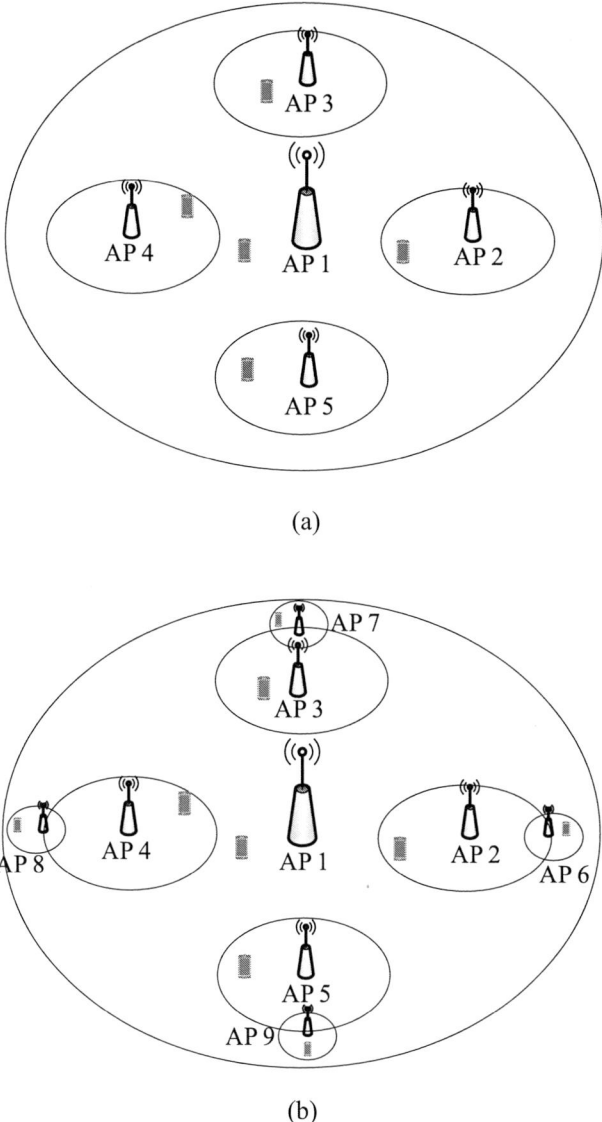

Fig. 3.5 Simulation model: (a) two-layer HetNet scenario; (b) three-layer HetNet scenario

maximum transmit power, and the maximum iteration number of them is set to 500. In particular, an early stop of iterations will happen if the difference of the sum-rates per link between two successive iterations is smaller than 0.001.

Figure 3.6 demonstrates the average sum-rate performance comparison in the two-layer HetNet scenario in the training stage. In this simulation, we set the channel correlation factor to be 0. This means that the channels in different time slots are

Table 3.1 Hyperparameters of each local DNN

Layers	$L_1^{(a)}$	$L_2^{(a)}$	$L_3^{(a)}$	$L_4^{(a)}$	$L_5^{(a)}$
Neuron number	7	100	100	1	1
Activation function	Linear	Relu	Relu	Sigmoid	Linear
Action noise ζ	Normal distribution with zero mean and variance 2				

Table 3.2 Hyperparameters of each actor DNN

Layers	$L_1^{(a)}$	$L_2^{(a)}$	$L_3^{(a)}$	$L_4^{(a)}$	$L_5^{(a)}$
Neuron number	7	100	100	1	1
Activation function	Linear	Relu	Relu	Sigmoid	Linear
Optimizer	Adam optimizer with learning rate 0.0001				
Mini-batch size D	128				
Learning rate $\tau_n^{(a)}$	$\tau_n^{(a)} = 0.001$				

Table 3.3 Hyperparameters of the critic DNN

Layers	$L_1^{(S)}$	$L_2^{(S)}$	$L_3^{(S)}$	$L_1^{(A)}$	$L_2^{(A)}$	$L_2^{(M)}$	$L_3^{(M)}$
Neuron number	$7N + N^2$	200	200	N	200	200	1
Activation function	Linear	Relu	Linear	Linear	Linear	Relu	Linear
Optimizer	Adam optimizer with learning rate 0.001						
Mini-batch size D	128						
Learning rate $\tau^{(c)}$	$\tau^{(c)} = 0.001$						
Discount factor η	0.5						

in IID. In the beginning, the proposed algorithm has a comparable sum-rate with the random power algorithm, and increases rapidly in the following time slots. This is because the proposed algorithm first goes through the random experience accumulation stage, where the transmit power of each AP is determined randomly. Then, only when the core network starts to train all actor DNNs and updates the trained weight parameters to each local DNN in AP can the sum-rate of the proposed algorithm increase. At around time slot 500, the proposed algorithm outperforms both WMMSE algorithm and FP algorithm. At around time slot 1500, the proposed algorithm converges and outperforms the benchmarks greatly. This indicates that both WMMSE algorithm and FP algorithm can only output sub-optimal solutions of the power control problem in each single time slot, resulting in the sub-optimal sum-rate performance. In comparison, by continuously exploring the different power control strategies with historical global information, the proposed algorithm can grasp the optimal power control strategy more accurately at each time slot, thus, improving the achievable sum-rate.

Figure 3.7 provides the corresponding sum-rate performance comparison in the two-layer HetNet in the testing stage. These CDF curves demonstrate the effectiveness of the proposed algorithm. Specifically, the proposed algorithm outperforms other benchmarks significantly, and is close to the performance of the upper bound. In particular, the sum-rate of the proposed algorithm in the testing stage is higher

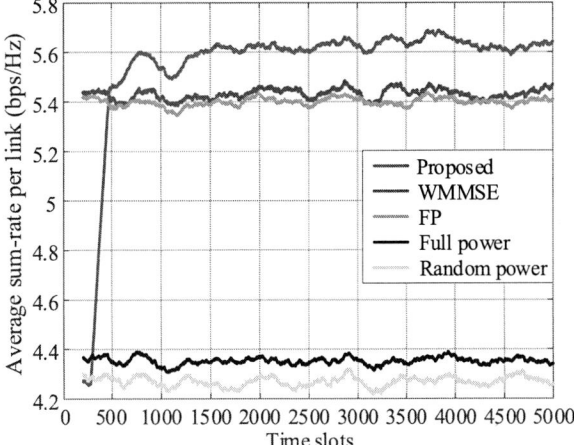

Fig. 3.6 Average sum-rate performance of the two-layer HetNet scenario in the training stage. The channel correlation factor is set to be zero, i.e., IID channel, and each value is the moving average of the previous 200 time slots

Fig. 3.7 Sum-rate performance of the two-layer intersecting scenario in the testing stage. The channel correlation factor is set to be zero, i.e., IID channel, and each value is the moving average of the previous 200 time slots

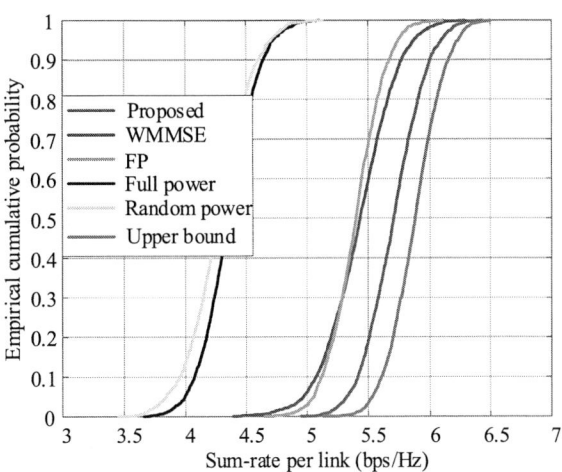

than that in the training stage. The reason is that in the training stage, the proposed algorithm explores the optimal power control policy at each time slot and may degrade the sum-rate performance after the proposed algorithm is already trained to convergence. However, in the testing stage, the proposed algorithm stops exploring and makes decisions with only the well-trained policy. This guarantees the full potential for optimizing the sum-rate of the proposed algorithm.

Figures 3.8 and 3.9 provide the average sum-rate performance comparison in the three-layer HetNet scenario in the training stage and the testing stage, respectively. Similarly to two-layer HetNet, we set the channel correlation factor to be 0, i.e., IID channel. The results in this simulation have the same trend as the two-layer

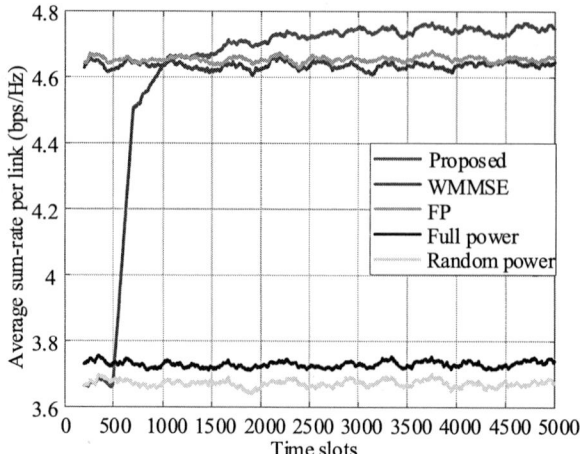

Fig. 3.8 Sum-rate performance of the three-layer HetNet scenario in the training stage. The channel correlation factor is set to be zero, i.e., IID channel, and each value is the moving average of the previous 200 time slots

Fig. 3.9 Sum-rate performance in the two-layer overlapping scenario in the testing stage. The channel correlation factor is set to be zero, i.e., IID channel, and each value is the moving average of the previous 200 time slots

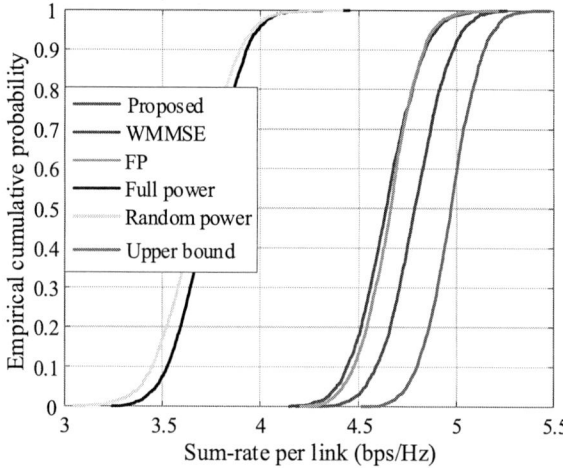

HetNet, i.e., the proposed algorithm outperforms other benchmarks at around time slot 1000, and converges at round time slot 3000. Additionally, Fig. 3.9 shows that the proposed algorithm can greatly exceed other benchmarks similar to Fig. 3.7.

Figure 3.10 provides the sum-rate performance comparison in the two-layer HetNet scenario with a random channel correlation factor ρ. In Fig. 3.10a, in the training stage, the proposed algorithm outperforms other benchmarks at around time slot 1000, and converges at around time slot 2000. In Fig. 3.10b, the proposed algorithm outperforms other benchmarks in most situations in the testing stage. These figures certify that the proposed algorithm stays effective even with random values of channel correlation factor ρ.

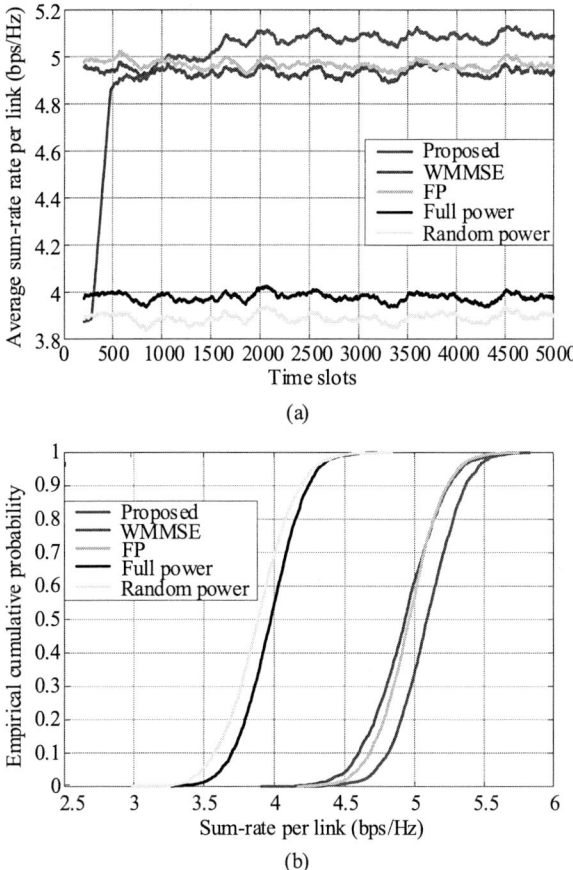

Fig. 3.10 Sum-rate performance of the two-layer HetNet scenario with a random channel correlation factor ρ. (**a**) Average sum-rate performance in the training stage. (**b**) Sum-rate performance in the testing stage

Similarly, Fig. 3.11 provides the sum-rate performance comparison in the three-layer HetNet scenario with a random channel correlation factor ρ. In Fig. 3.11a, the proposed algorithm in the training stage increases and converges at around 3000 time slots. Different from the simulation results shown above, the proposed algorithm can only achieve a comparable sum-rate with the WMMSE algorithm and FP algorithm after convergence. In Fig. 3.11b, in the testing stage, the sum-rate performance of the proposed algorithm is almost the same as the WMMSE algorithm and FP algorithm. That is because, it is challenging for the core network to learn an optimal power control policy as the network size increases. To this end, there is a degradation in the sum-rate performance of the proposed algorithm compared to the two-layer HetNet scenario.

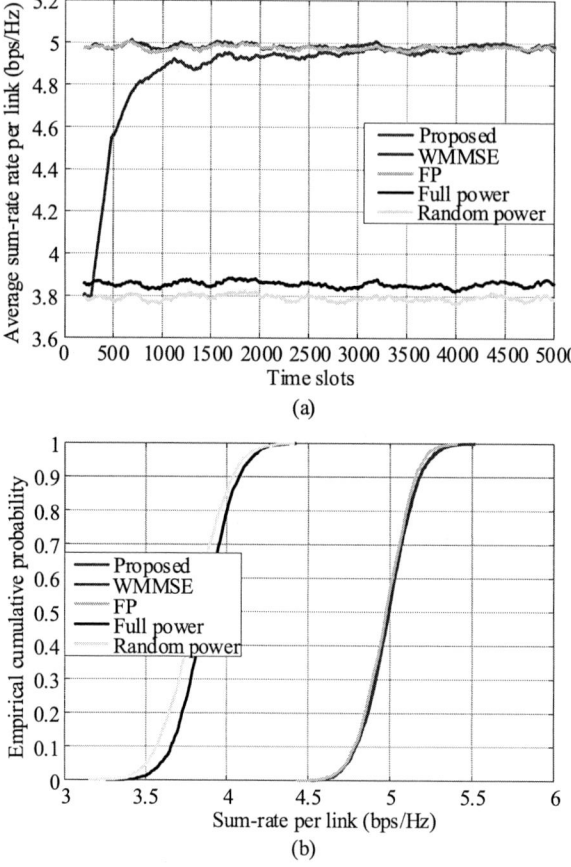

Fig. 3.11 Sum-rate performance in the three-layer HetNet scenario with a random channel correlation factor ρ. (**a**) Average sum-rate performance in the training stage. (**b**) Sum-rate performance in the testing stage

Table 3.4 demonstrates the average time complexity comparison. In the training stage, the proposed algorithm trains a critic DNN and N actor DNN in total. The training time of the critic DNN is 9.7 ms and the training time of each actor DNN is 5.9 ms. In the testing stage, the proposed algorithm only consumes 0.34 ms to determine a proper transmit power. In comparison, the execution time of the WMMSE algorithm and FP algorithm is much longer than the proposed algorithm, i.e., 120 and 79 ms, respectively. It should be noted that the training and testing time is highly dependent on the computational capacity of the equipment. Therefore, the time complexity can be further reduced in practical situations with stronger computational equipment.

Table 3.4 Average time complexity

Training the critic DNN	Training an actor DNN	Calculation with a local DNN	WMMSE	FP
9.7 ms	5.9 ms	0.34 ms	120 ms	79 ms

3.6 Conclusions

In this chapter, we leverage the DRL to design a multi-agent power control with centralized learning for a HetNet, which will be a main component of the wireless edge network. We established a local DNN at each AP and proposed MASC method to train all local DNNs in the cloud in a centralized way. To this end, each AP can determine the optimal transmit power utilizing the local DNN with local observations. Simulation results show that the proposed algorithm can both enhance the sum-rate and reduce the time complexity compared to the benchmarks, e.g., the WMMSE algorithm and the FP algorithm.

References

1. Zhang, L., Liang, Y.-C., Niyato, D.: 6G visions: mobile ultra-broadband, super Internet-of-Things, and artificial intelligence. China Commun. **16**(8), 1–14 (2019)
2. Andrews, J.G., Buzzi, S., Choi, W., Hanly, S.V., Lozano, A., Soong, A.C.K., Zhang, J.C.: What will 5G be? IEEE J. Select. Areas Commun. **32**(6), 1065–1082 (2014)
3. Zhang, L., Xiao, M., Wu, G., Alam, M., Liang, Y.-C., Li, S.: A survey of advanced techniques for spectrum sharing in 5G networks. IEEE Wirel. Commun. **24**(5), 44–51 (2017)
4. Yang, C., Li, J., Guizani, M., Anpalagan, A., Elkashlan, M.: Advanced spectrum sharing in 5G cognitive heterogeneous networks. IEEE Wirel. Commun. **23**(2), 94–101 (2016)
5. Singh, S., Andrews, J.G.: Joint resource partitioning and offloading in heterogeneous cellular networks. IEEE Wirel. Commun. **13**(2), 888–901 (2014)
6. Shi, Q., Razaviyayn, M., Luo, Z.-Q., He, C.: An iteratively weighted MMSE approach to distributed sum-utility maximization for a MIMO interfering broadcast channel. IEEE Trans. Signal Process. **59**(9), 4331–4340 (2011)
7. Shen, K., Yu, W.: Fractional programming for communication systems-Part I: power control and beamforming. IEEE Trans. Signal Process. **66**(10), 2616–2630 (2018)
8. Hu, R.Q., Qian, Y.: An energy efficient and spectrum efficient wireless heterogeneous network framework for 5G systems. IEEE Commun. Mag. **52**(5), 94–101 (2014)
9. Huang, J., Berry, R.A., Honig, M.L.: Distributed interference compensation for wireless networks. IEEE J. Sel. Areas Commun. **24**(5), 1074–1084 (2006)
10. Zhang, H., Venturino, L., Prasad, N., Li, P., Rangarajan, S., Wang, X.: Weighted sum-rate maximization in multi-cell networks via coordinated scheduling and discrete power control. IEEE J. Sel. Areas Commun. **29**(6), 1214–1224 (2011)
11. Le, L.B., Hossain, E.: Resource allocation for spectrum underlay in cognitive radio networks. IEEE Trans. Wirel. Commun. **7**(12), 5306–5315 (2008)
12. Nasir, Y.S., Guo, D.: Multi-agent deep reinforcement learning for dynamic power allocation in wireless networks. IEEE J. Sel. Areas Commun. **37**(10), 2239–2250 (2019)
13. Xiao, L., Zhang, H., Xiao, Y., Wan, X., Liu, S., Wang, L.-C., Poor, H.V.: Reinforcement learning-based downlink interference control for ultra-dense small cells. IEEE Trans. Wirel. Commun. **19**(1), 423–434 (2020)

14. Amiri, R., Almasi, M.A., Andrews, J.G., Mehrpouyan, H.: Reinforcement learning for self organization and power Control of two-tier heterogeneous networks. IEEE Trans. Wirel. Commun. **18**(8), 3933–3947 (2019)
15. Sun, Y., Feng, G., Qin, S., Liang, Y.-C., Yum, T.-S.P.: The SMART handoff policy for millimeter wave heterogeneous cellular networks. IEEE Trans. Mobile Commun. **17**(6), 1456–1468 (2018)
16. Nguyen, D.D., Nguyen, H.X., White, L.B.: Reinforcement learning with network-assisted feedback for heterogeneous RAT selection. IEEE Trans. Wirel. Commun. **16**(9), 6062–6076 (2017)
17. Wei, Y., Yu, F.R., Song, M., Han, Z.: User scheduling and resource allocation in HetNets with hybrid energy supply: an actor-critic reinforcement learning approach. IEEE Trans. Wirel. Commun. **17**(1), 680–692 (2018)
18. Morozs, N., Clarke, T., Grace, D.: Heuristically accelerated reinforcement learning for dynamic secondary spectrum sharing. IEEE Access **3**, 2771–2783 (2015)
19. Raj, V., Dias, I., Tholeti, T., Kalyani, S.: Spectrum access in cognitive radio using a two-stage reinforcement learning approach. IEEE J. Sel. Topics Signal Process. **12**(1), 20–34 (2018)
20. Iacoboaiea, O., Sayrac, B., Jemaa, S.B., Bianchi, P.: SON coordination in heterogeneous networks: a reinforcement learning framework. IEEE Trans. Wirel. Commun. **15**(9), 5835–5847 (2016)
21. Mnih, V., et al.: Human-level control through deep reinforcement learning. Nature **518**(7540), 529–533 (2015)
22. Zhang, C., Patras, P., Haddadi, H.: Deep learning in mobile and wireless networking: a survey. IEEE Commun. Surv. Tuts. **21**(3), 2224–2287 (2019)
23. Chien, T.V., Canh, T.N., Bjornson, E., Larsson, E.G.: Power control in cellular massive MIMO with varying user activity: a deep learning solution. IEEE Trans. Wirel. Commun. **19**(9), 5732–5748 (2020). https://doi.org/10.1109/TWC.2020.2996368
24. Mennes, R., De Figueiredo, F.A.P., Latre, S.: Multi-agent deep learning for multi-channel access in slotted wireless networks. IEEE Access **8**, 95032–95045 (2020)
25. Cui, W., Shen, K., Yu, W.: Spatial deep learning for wireless scheduling. IEEE J. Select. Areas Commun. **37**(6), 1248–1261 (2019)
26. Ye, H., Li, G.Y., Juang, B.-H.F.: Deep reinforcement learning based resource allocation for V2V communications. IEEE Trans. Veh. Technol. **68**(4), 3163–3173 (2019)
27. Yu, Y., Wang, T., Liew, S.C.: Deep-reinforcement learning multiple access for heterogeneous wireless networks. IEEE J. Sel. Areas Commun. **37**(6), 1277–1290 (2019)
28. He, Y., Zhang, Z., Yu, F.R., Zhao, N., Yin, H., Leung, V.C.M., Zhang, Y.: Deep-reinforcement-learning-based optimization for cache-enabled opportunistic interference alignment wireless networks. IEEE Trans. Veh. Technol. **66**(11), 10433–10445 (2017)
29. Zhang, L., Tan, J., Liang, Y.-C., Feng, G., Niyato, D.: Deep reinforcement learning-Based modulation and coding scheme selection in cognitive heterogeneous networks. IEEE Trans. Wirel. Commun. **18**(6), 3281–3294 (2019)
30. Mismar, F.B., Evans, B.L., Alkhateeb, A.: Deep reinforcement learning for 5G networks: Joint beamforming, power control, and interference coordination. IEEE Trans. Commun. **68**(3), 1581–1592 (2020)
31. Huang, H., Yang, Y., Wang, H., Ding, Z., Sari, H., Adachi, F.: Deep reinforcement learning for UAV navigation through massive MIMO technique. IEEE Trans. Veh. Technol. **69**(1), 1117–1121 (2020)
32. Zhang, H., Yang, N., Huangfu, W., Long, K., Leung, V.C.M.: Power control based on deep reinforcement learning for spectrum sharing. IEEE Trans. Wirel. Commun. **19**(6), 4209–4219 (2020)
33. Luong, N.C., et al.: Applications of deep reinforcement learning in communications and networking: A survey," IEEE Commun. Surveys, vol. 21, no. 4, pp. 3133–3174, 2019.
34. Kim, T., Love, D.J., Clerckx, B.: Does frequent low resolution feedback outperform infrequent high resolution feedback for multiple antenna beamforming systems? IEEE Trans. Signal Process. **59**(4), 1654–1669 (2011)

35. Luo, Z.-Q., Zhang, S.: Dynamic spectrum management: complexity and duality. IEEE J. Sel. Topics Signal Process. **2**(1), 57–73 (2008)
36. ETSI TR 125 942: Universal Mobile Telecommunications System (UMTS); RF system scenarios (3GPP TR 25.942 version 14.0.0), Apr. 2017

Chapter 4
AI-Enabled Distributed Spectrum Sharing

Abstract This chapter investigates the AI-enabled distributed spectrum sharing in the wireless edge network, where multiple BSs share the same bandwidth and may suffer severe interference. This inevitably degrades the network spectrum efficiency (SE). To tackle this issue, conventional methods involve centralized power control strategies that rely on global instantaneous CSI. In this chapter, we propose a multi-agent independent actor-critic (MAIAC) power control algorithm for each BS to optimize local transmit power by enabling lightweight collaborations between the core network (cloud) and different BSs (edge). In particular, each BS learns its the power control policy locally with a cloud-based global reward mechanism. Simulation results show that the proposed MAIAC algorithms can respectively achieve over 99% SE performance of conventional algorithms in quasi-static scenarios, and approximately 89 $\sim$ 100% in dynamic scenarios with significantly reduced time complexity.

Keywords Cloud-edge collaboration · Spectrum efficiency · Energy efficiency · Power control · DRL · HetNet

4.1 Introduction

In recent years, with the rapid developments of mobile communication technology and the continuous proliferation of smart devices, mobile data traffic has experienced unprecedented growth. The widespread use of smartphones, tablets, and other IoT devices has fueled a growing demand for high-speed, high-capacity data services among users [1, 2]. This immense volume of data poses significant challenges for existing cellular systems.

To tackle this issue, HetNet has emerged as an effective solution in the wireless edge network that allows different types of small BSs to coexist with the macro BSs by sharing the same bandwidth [3, 4]. Nevertheless, the shared bandwidth for these BSs will inevitably bring about severe interference, which significantly influences the communication quality of the whole HetNet. To evaluate how effectively a network utilizes its allocated spectrum, *spectrum efficiency* (SE), referring to the

L. Zhang et al., *AI-enabled Spectrum Sharing*, SpringerBriefs in Computer Science,
https://doi.org/10.1007/978-981-97-7644-3_4

amount of data that can be transmitted per unit of frequency bandwidth in wireless communication, is widely studied in the research of wireless communication systems [5, 6]. Indeed, improving SE means maximizing the utilization of the available spectrum resources, which is especially important in crowded urban areas where the available spectrum is limited. Consequently, the pursuit of improving SE has gained significant global attention in recent years.

In a typical underlay HetNets, the optimization of joint power plays a crucial role in achieving desirable SE performance. As a result, numerous studies have been conducted in recent years. For example, the WMMSE algorithm was introduced in [7] for the weighted sum rate maximization subject to a power constraint, and it can converge to a local optimal solution. The FP algorithm was proposed in [8], which applied a closed-form updating process for the power control problem and achieved a near-optimal weighted sum rate performance.

Over the last few years, researchers have explored the potential of AI techniques to enhance SE in wireless communication systems. Several studies have successfully employed DL methods to address the SE maximization problem. For instance, in [9], a *convolutional neural network* (CNN) was used to extract spatial features from the channel gain matrix, which was then fed into a transmit power optimizer to minimize the SE or energy efficiency loss. Zhou et al. [10] focused on optimizing transmit power in *cognitive radio* (CR) networks using a DNN.

RL is another widely used AI technique in the wireless communication field. RL agents operate in an online manner by continuously taking actions according to a policy, receiving rewards from the environment, transitioning to new environment states, and updating action policy. In this way, RL agents can converge to an optimal action policy that maximizes rewards. DRL is an effective and popular technique that combines DL and RL. This technique has been successfully employed to tackle complex problems with large state and action spaces [11–14]. Further, DRL has been applied successfully in many scenarios [15–19]. [16] utilized DDPG algorithm in DRL to optimize the sum rate of the network, which is comparable to the performance of WMMSE and FP algorithms. Iqbal et al. [18] and Liu et al. [19] adopt DQN variants to respectively optimize SE and energy efficiency while satisfying the user requirements. The results show that the proposed DRL algorithm can outperform the existing DRL schemes.

As discussed above, conventional optimization algorithms, e.g., [7] and [8], can achieve high SE performance under the premise of the collection of global instantaneous CSI of the whole network, which is challenging to obtain in practical scenarios. Besides, they typically have high computation latency and may make the optimization result outdated. Existing AI-based power control optimization methods can be roughly divided into two categories, i.e., DL-based algorithms and DRL-based algorithms. The DL-based optimization algorithms, e.g., [20], also need to collect global instantaneous CSI in advance for both training and testing, meanwhile, DL methods may encounter the challenge of inconsistency between offline training datasets and online testing datasets, particularly in rapidly changing HetNet environments. The DRL-based optimization algorithms, e.g., [21–23], can achieve high SE performance through massive information exchange among the core network and different BSs.

To deal with the issues above, we propose an intelligent power control algorithm in a HetNet, with which each BS can control its transmit power to enhance the global SE in a distributed manner. On the one hand, we enable lightweight collaborations among core network (cloud) and different BSs (edge), and develop a lightweight cloud-edge collaboration framework to enhance global SE. The lightweight collaboration lies in that, the edge BSs do not need to exchange local instantaneous information with each other, and the cloud collects only historical data rate from edge BSs and calculates global historical SE reward. On the other hand, we establish an independent actor-critic structure for each BS, and develop a multi-agent independent actor-critic (MAIAC) power control algorithm with distributed learning, such that each BS can learn and decide local power control policy to enhance the SE in a distributed manner.

4.2 Syetem Model

We consider a HetNet that comprises a macro BS and N micro BSs, as shown in Fig. 4.1. Each BS is responsible for serving a corresponding *user equipment* (UE), and both the BSs and UEs implement a single antenna. The UE served by BS n is denoted as UE n, where $n = \{0, 1, 2, \cdots, N\}$ and index 0 represents the macro BS and its served UE. Note that all the BSs share the same spectrum band for synchronized downlink transmissions.

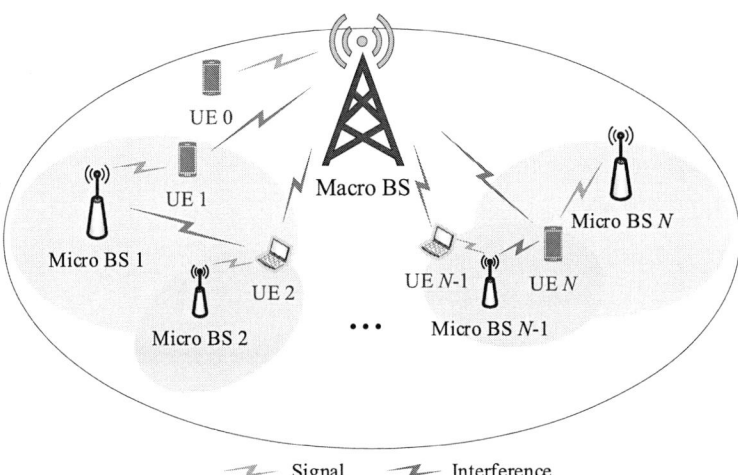

Fig. 4.1 A typical HetNet

4.2.1   *Channel and Signal Models*

We model the wireless channel as being composed of two components: large-scale attenuation (path-loss and log-normal shadowing) and small-scale Rayleigh fading. The path-loss is determined by the distance between the BS and the UE, while the shadowing typically remains the same over many time slots. We can represent the large-scale attenuation from BS n to UE k as the sum of path-loss $L_{n \to k}$ and shadowing $\phi_{n \to k}$, denoted as $\chi_{n \to k} = L_{n \to k} + \phi_{n \to k}$. We model $\phi_{n \to k}$ in time slot t as [24]

$$\phi_{n \to k}(t) = \rho_s \cdot \phi_{n \to k}(t-1) + \sqrt{1 - \rho_s^2} \cdot N_s(t), \tag{4.1}$$

where $N_s(t)$ is a *independent and identically distributed* (IID) Gaussian variables with 0 mean and lognormal shadowing standard deviation σ_s. The correlation factor ρ_s is computed by $\rho_s = 2^{-\Delta d_n / d_{\text{corr}}}$, where Δd_n is the displacement of UE n between two successive time slots, and d_{corr} is the decorrelation length of the environment.

The Rayleigh fading from BS n to UE k is denoted as $h_{n \to k}$, which is modeled as a Markov process using the Jake's fading model [25], i.e.,

$$h_{n \to k}(t) = \rho_R \cdot h_{n \to k}(t-1) + \kappa_{n \to k}, \tag{4.2}$$

where $h_{n \to k}(0) \sim \mathcal{CN}(0, 1)$ is a complex Gaussian distribution, ρ_R is the correlation coefficient and $\kappa_{n \to k}$ is a random variable with distribution $\mathcal{CN}(0, 1 - \rho_R^2)$.

Then, the combined downlink channel gain can be formulated as $g_{n \to k} = \chi_{n \to k} |h_{n \to k}|^2$. Meanwhile, the received signal at UE n includes the desired signal of UE n, the interference, and the Gaussian noise with power σ^2 denoted by n_0, i.e.,

$$y_n(t) = \sqrt{p_n(t)\chi_{n \to n}} h_{n \to n}(t)x_n(t) + \sum_{k \in N, k \neq n} \sqrt{p_k(t)\chi_{k \to n}} h_{k \to n}(t)x_k(t) + n_0, \tag{4.3}$$

where $p_n(t)$ is the transmit power of BS n, $x_n(t)$ is the transmitted signal with unit power. The SINR at UE n can hence be written as

$$\gamma_n(t) = \frac{p_n(t)g_{n \to n}(t)}{\sum_{k \in N, k \neq n} p_k(t)g_{k \to n}(t) + \sigma^2}. \tag{4.4}$$

Let B denote the available bandwidth, then the achievable data rate of UE n can be calculated based on the Shannon formula, i.e.,

$$r_n(t) = B \log\left(1 + \gamma_n(t)\right). \tag{4.5}$$

Then, the normalized global SE in time slot t can be expressed as

$$SE(t) = \frac{\sum_{n=0}^{N} r_n(t)}{B}.$$

(4.6)

As discussed above, we formulate the SE maximization problem in each time slot as:

$$\text{s.t.}\quad \max_{p_n(t), \forall n \in N} SE(t) = \frac{\sum_{n=0}^{N} r_n(t)}{B}$$

$$0 \le p_n(t) \le p_{n,\max}, \forall n \in N,$$

(4.7)

where $p_{n,\max}$ is the maximum transmit power of BS n. Note that in a common HetNet, different BS are deployed in various areas to provide services to UEs. Due to differing physical characteristics, coverage requirements, and interference control factors, different types of BS can have distinct maximum transmit power constraints. For instance, macro BS generally has higher maximum transmit power constraints to cover larger areas and serve more UEs. In contrast, micro BS usually have lower maximum transmit power constraints because they are primarily used to provide localized coverage and enhance network capacity. Therefore, based on these varying constraints, power control strategies need to be adjusted according to the specific constraints of each BS.

On the other hand, the optimization of (4.7) is known to be NP-hard with high time complexity [26]. To overcome these issues, we provide a lightweight cloud-edge collaboration framework along with an MAIAC power control algorithm, which is elaborated in Sects. 4.3 and 4.4, respectively.

4.3   Lightweight Cloud-Edge Collaboration Framework

From (4.7), it is evident that the optimization of global SE in the HetNet is highly related to global instantaneous information including the achievable data rate of all the BSs. This global instantaneous information is critical for achieving optimal global SE performance. However, constructing this global instantaneous information requires each BS to collect not only its own local instantaneous information, but also the non-local instantaneous information from other BSs, which requires massive information exchange among local BSs (if possible), leading to significant overheads.

We observe that, it is easy for the cloud to collect historical local SE related data from each edge BS, which contains useful information on the optimal power control policy of each edge BS. Thus, it is possible to design lightweight collaborations among core network (cloud) and different BSs (edge). With the lightweight collaborations, the edge BSs do not need to exchange any instantaneous information with each other, and the cloud collects only historical local SE related data from each edge BS and calculates historical global SE reward, which will then be fed back to each BS to improve the corresponding local power control policy. On the other hand, the conventional methods for optimizing SE involve modeling different optimization problems and applying different optimization theories, meaning a complicated design process.

Based on the above observations, we develop a lightweight cloud-edge collaboration framework with lightweight collaborations among the cloud and different BSs, as illustrated in Fig. 4.2. In the edge part, each edge BS is viewed as an agent with an actor DNN responsible for determining its local transmit power. Meanwhile, each edge BS is equipped with a critic DNN and two experience replay buffers, namely, local experience replay buffer and local-global experience replay buffer. In particular, the critic DNN is designed to guide the periodic training of the corresponding actor DNN. The local experience replay buffer stores the local immediate experiences of each BS, while the local-global experience replay buffer stores the integration of local experience and global historical SE reward received from the cloud. The cloud is comprised of a group of queues (a receive queue

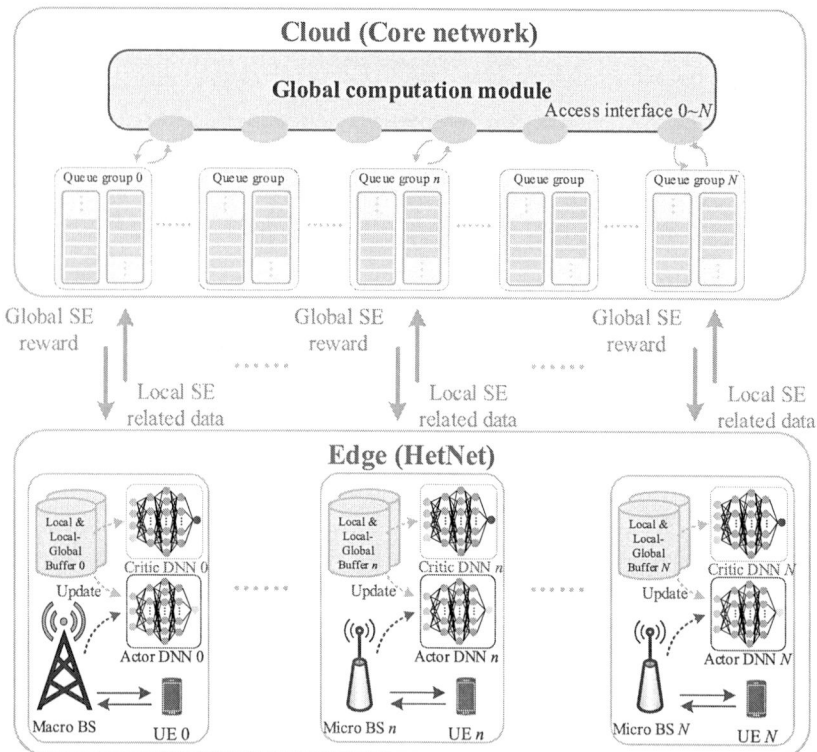

Fig. 4.2 Lightweight cloud-edge collaboration framework

and a transmit queue) for exchanging historical information with each BS, as well as a global computation module. Each edge BS is connected to the cloud via a bidirectional wired or wireless link with significant latency.

At the beginning of each time slot, each BS computes its transmit power by inputting local information into the actor DNN. At the end of each time slot, each BS stores the corresponding local experience and timestamp into its local experience replay buffer, and uploads only SE related data and the timestamp to the cloud. Upon receiving the data from each edge BS, the cloud stores the data in its receive queue and schedules them to calculate global historical SE reward using its global computation module. The calculated global SE reward together with the corresponding timestamp is placed in each transmit queue, and will be transmitted to each edge BS in a first-in-first-out manner. Upon receiving the data from the cloud, each BS retrieves the local experience with the same timestamp from its local experience replay buffer, and combines it with the global SE reward to form a local-global experience, which is stored in the local-global experience replay buffer. In this way, each edge BS can sample mini-batches of local-global experience to train periodically both actor DNN and critic DNN. This process will be terminated until the convergence of each actor DNN. To this end, each actor DNN has learned a local power control policy to enhance the global SE.

The proposed framework shows notable features. Firstly, compared to conventional power control optimization algorithms, the proposed framework enables each BS to determine an appropriate transmit power using only local information. This ensures real-time power control and reduces computation caused by centralized optimization. Secondly, the proposed framework enables lightweight collaborations among the cloud and different edge BSs, which greatly minimizes the information exchange overheads.

4.4 Spectrum-Efficient Multi-Agent Independent Actor-Critic Power Control Algorithm

In this section, we will introduce spectrum-efficient MAIAC power control algorithm, which is built upon the framework presented in the previous section, enabling each edge BS to learn and optimize their local power control policies to enhance the SE. In the following, we elaborate the designs of each algorithm, including state and action representations, experience acquisition, global rewards, DNN structures, and DNN training methods.

4.4.1 Designs of State and Action

At time slot t, the state of the actor DNN n is composed of the historical local information in the previous time slot, the local instantaneous information in the current time slot, i.e.,

- Historical local information: The historical local information comprises the channel gain denoted by $g_{n \to n}(t-1)$, the transmit power of edge BS n denoted by $p_n(t-1)$, the received interference denoted by $\sum_{k \in N, k \neq n} p_k(t-1)g_{k \to n}(t-1)$, the received SINR denoted by $\gamma_n(t-1)$, and the data rate denoted by $r_n(t-1)$, which has a total of five elements.
- Local instantaneous information: The local instantaneous information comprises the channel gain denoted by $g_{n \to n}(t)$, and the received interference before changing transmit power denoted by $\sum_{k \in N, k \neq n} p_k(t-1)g_{k \to n}(t)$, which has a total of two elements.

The state of actor DNN n for SE optimization problem is denoted as s_n. To construct s_n at time slot t, we incorporate both the historical local information and the local instantaneous information, which can be expressed as:

$$s_n(t) = \left\{ g_{n \to n}(t-1), \; p_n(t-1), \; \sum_{k \in N, k \neq n} p_k(t-1)g_{k \to n}(t-1), \right.$$

$$\left. \gamma_n(t-1), \; r_n(t-1), \; g_{n \to n}(t), \; \sum_{k \in N, k \neq n} p_k(t-1)g_{k \to n}(t) \right\}. \tag{4.8}$$

Given that each BS is responsible for determining its transmit power and optimizing the SE performance, the action that the actor DNN n takes (denoted as a_n) is designed as its transmit power at the current time slot.

4.4.2 Designs of Experience

As discussed in Sect. 4.3, each BS is equipped with both a local experience replay buffer and a local-global experience replay buffer. At the end of time slot t, the local experience of BS n for SE optimization problem can be obtained and constructed as:

$$e_n^l(t) = \{s_n(t-1), a_n(t-1), s_n(t)\}. \tag{4.9}$$

After receiving the global historical SE reward from the cloud, the local-global experience in the BS can be constructed by integrating the corresponding local experience with the global reward. Let us denote the global SE reward as R, and the transmission latency is represented as T_l. Under these considerations, we can express the local-global experience of BS n for SE optimization problem as follows:

$$e_n^{l-g}(t) = \{s_n(t-1), a_n(t-1), R(t-1+T_l), s_n(t)\}. \tag{4.10}$$

4.4.3 Designs of Global SE Reward

The global reward is designed to evaluate the effectiveness of the joint action based on the observed state. In particular, if we aim to optimize the global SE, the global reward in time slot t is designed as:

$$R(t) = \text{SE}(t - T_l). \tag{4.11}$$

Subsequently, to calculate the global SE reward, each edge BS is required to upload its local SE related data. Referring to (4.7), the local SE related data of BS n is formulated as its local data rate, i.e.,

$$L = r_n. \tag{4.12}$$

In this manner, the cloud will collect local SE related data L, along with the current timestamp t_i, and upload them to the cloud at the end of each time slot. Note that, the size of the transmitted data maintains constant space complexity, which significantly reduces the signaling exchange overheads and ensures lightweight collaborations between the cloud and the edge.

4.4.4 Designs of DNNs' Structure

We establish an independent actor-critic structure for each BS to ensure real-time power control using only local information. The network structure of actor DNN n is shown in Fig. 4.3. Specifically, it inputs seven neurons corresponding to the elements in s_n. These input ports will be then fed to several fully-connected hidden layers. The output layer produces the action (a_n), which corresponds to the transmit power. This action is scaled within the range of 0 to $p_{n,max}$.

The critic DNNs are designed to evaluate the global SE, as shown in Fig. 4.4. The critic DNN n comprises three fully-connected NN modules, i.e., the state module, the action module, and the value evaluation module, respectively. The state module receives the local state s_n as input, while the action module receives the local action a_n as input, and the value evaluation module receives the output layers of both modules as input, which then outputs the Q value.

Fig. 4.3 The network
structure of actor DNN n

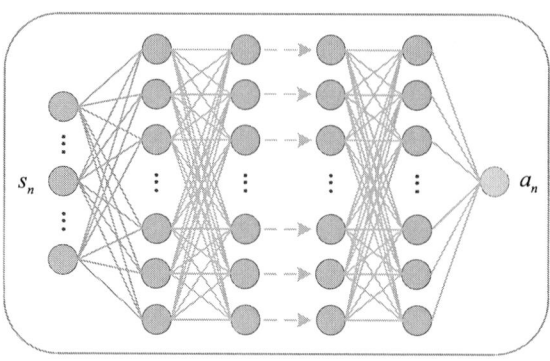

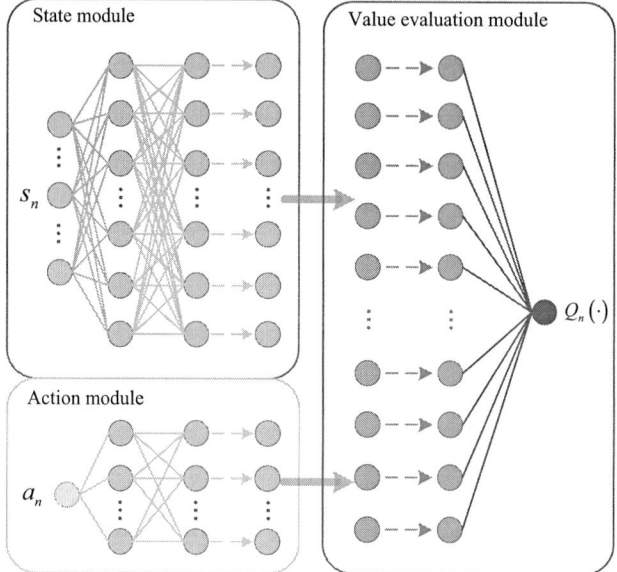

Fig. 4.4 The network structure of critic DNN n

4.4.5 Spectrum-Efficient DNN Training Methods

In this part, we propose a spectrum-efficient MAIAC power control algorithm based
on the policy gradient-based DDPG algorithm in [27], as shown in Algorithm 1.

Initially, a random transmit power is selected by all the BSs for their downlink
transmissions. By continuously exchanging local related data and historical global
rewards with the cloud, the BSs are able to construct and store local-global experiences in their local-global experience replay buffers. Upon reaching a minimum of
D experiences, each BS will sample the experiences to compute the gradients and
train the DNNs.

In the following, we denote actor DNN n and critic DNN n as $\mu_n^{(a)}\left(s_n; \omega_n^{(a)}\right)$ and $Q_n^{(c)}\left(s_n, a_n; \omega_n^{(c)}\right)$, respectively, where $\omega_n^{(a)}$ and $\omega_n^{(c)}$ represent the parameters of actor DNN n and critic DNN n, respectively. Besides, we create target critic DNN n denoted by $Q_n^{(c\text{-})}\left(s_n, a_n; \omega_n^{(c\text{-})}\right)$, and target actor DNN n denoted by $\mu_n^{(a\text{-})}\left(s_n; \omega_n^{(a\text{-})}\right)$, which helps stabilize the training of the corresponding critic DNN n and actor DNN n. In the following, we provide the training procedures for the critic, actor and target DNNs, respectively.

For edge BS n, the long-term target $y_n^{(\text{Tar})}$ of critic DNN n can be derived by

$$y_n^{(\text{Tar})} = R + \eta Q_n^{(c\text{-})}\left(s_n', a_n'; \omega_n^{(c\text{-})}\right), \tag{4.13}$$

where η is the discount factor, a_n' is the output of $\mu_n^{(a\text{-})}\left(s_n'; \omega_n^{(a\text{-})}\right)$. Then, the loss function of critic DNN n is equal to the *Mean Squared Error* (MSE) between the predicted value $Q_n^{(c)}(s_n, a_n; \omega_n^{(c)})$ and target value $y_n^{(\text{Tar})}$ in terms of the D samples, i.e.,

$$\mathbb{L}(\omega_n^{(c)}) = \frac{1}{D}\sum_i [y_n^{(\text{Tar})} - Q_n^{(c)}(s_n, a_n; \omega_n^{(c)})]^2. \tag{4.14}$$

Then, each critic DNN is capable of backpropagating the loss and updating its parameter $\omega_n^{(c)}$ through a gradient descent-based method. By repeating the above process, each critic DNN can gradually optimize its parameters towards the optimal SE performance. After that, each target critic DNN can be updated using a soft update method, i.e.,

$$\omega_n^{(c\text{-})} = \tau^{(c)}\omega_n^{(c)} + (1 - \tau^{(c)})\omega_n^{(c\text{-})}, \tag{4.15}$$

where $\tau^{(c)}$ is the soft update rate of each target critic DNN.

The actor DNNs are trained by utilizing gradients obtained from the critic DNNs. Specifically, the parameters of the actor DNN n is updated to maximize the Q value of critic DNN n, i.e., $\omega_n^{(a)} = \omega_n^{(a)} - \alpha^{(a)}\nabla_{\omega_n^{(a)}} J(\omega_n^{(a)})$, where $\alpha^{(a)}$ is the learning rate of actor DNN n, $\nabla_{\omega_n^{(a)}} J(\omega_n^{(a)})$ is the derivation of the average Q value to $\omega_n^{(a)}$, which can be formulated as:

$$\nabla_{\omega_n^{(a)}} J(\omega_n^{(a)}) \approx \frac{1}{D}\sum_i \nabla_{a_n} Q_n^{(c)}\left(s_n, a_n; \omega_n^{(c)}\right)$$

$$\Big|_{a_n = \mu_n^{(a)}(s_n; \omega_n^{(a)})} \nabla_{\omega_n^{(a)}} \mu_n^{(a)}(s_n; \omega_n^{(a)}). \tag{4.16}$$

Similarly, the parameters of each target actor DNN are updated as follows:

$$\omega_n^{(a-)} = \tau^{(a)} \omega_n^{(a)} + (1 - \tau^{(a)}) \omega_n^{(a-)}. \qquad (4.17)$$

where $\tau^{(a)}$ is the soft update rate of each target actor DNN.

To enable effective exploration and exploitation of the wireless environment, we add exploration noise to the actor policy, i.e.,

$$p_n = \mu_n^{(a)}(s_n; \omega_n^{(a)}) + \mathcal{N}_a, \qquad (4.18)$$

where $\mathcal{N}_a$ represents a Gaussian noise with a zero-mean and variance of ζ. The noise variance ζ is initialized with ζ_{ini}, decays exponentially at a fixed rate λ, and will remain as ζ_{min} when it is lower than ζ_{min}.

Algorithm 3 Proposed spectrum-efficient MAIAC power control algorithm.

1: **Initialization:**
2: Initialize actor DNN $\mu_n^{(a)}(s_n; \omega_n^{(a)})$, target actor DNN $\mu_n^{(a-)}(s_n'; \omega_n^{(a-)})$, critic DNN $Q_n^{(c)}(s_n, a_n; \omega_n^{(c)})$, and target critic DNN $Q_n^{(c-)}(s_n', a_n'; \omega_n^{(c-)})$ for all $n \in N$.
3: **Experience accumulation:**
4: At the beginning of each time slot, each BS selects a random transmit power to transmit data.
5: At the end of each time slot, each BS constructs local experience and stores it in its local experience buffer together with a timestamp. Then, each BS uploads only local SE related data L and the timestamp to the cloud.
6: Upon receipt, the cloud stores the data in its receive queue and schedules them for processing to calculate the global historical SE reward. The calculated results, along with the timestamp, will be first placed in each transmit queue and then transmitted to each edge BS.
7: Upon receiving the global SE reward, each BS retrieves the local experience with the same timestamp from its local experience replay buffer, and combines it with global SE reward to form the local-global experience, which is then stored in its local-global experience replay buffer.
8: Upon each BS having D local-global experiences, a batch of local-global experiences will be sampled to update each critic DNN and actor DNN, as well as each target critic DNN and target actor DNN.
9: **Repeat until convergence:**
10: At the beginning of time slot t, each BS utilizes its local actor DNN to determine a transmit power. After that, each BS constructs local experience, and uploads L to the cloud together with a timestamp.
11: In time slot $t + 2T_l$, BS n receives R from cloud, forms local-global experiences, and updates DNNs and their corresponding target DNNs.

4.5   Simulation Results

In this section, we conduct separate simulations to analyze the SE performance in both the quasi-static and dynamic scenarios.

We consider a two-layer HetNet scenario with a macro BS and four micro BSs. The macro BS is deployed in the first layer, located at coordinates $(0, 0)$, and provides coverage to a circular area with a radius between 10 and 1000 meters. Micro BS 1 to BS 4 are distributed in the second layer, located at coordinates $(500, 0)$, $(0, 500)$, $(-500, 0)$, $(0, -500)$, respectively. These micro BSs have coverage areas with radius ranging from 10 to 200 meters. Notably, the macro BS is subjected to a maximum transmit power of 30 dBm, while the micro BSs are subjected to a maximum transmit power of 23 dBm.

Besides, the parameters utilized in our simulation are presented in Table 4.1. In the following, we analyze the performance of the proposed MAIAC algorithms for two phases, namely, the training and testing phases. During the training phase, the proposed MAIAC algorithms are initialized and trained from scratch for 30,000 time slots. After that, the algorithms transition to the testing phase where it is evaluated

Table 4.1 Parameters settings

Parameter	Value
Bandwidth B	10 Mhz
Noise power σ^2	-114 dBm
correlation coefficient ρ_R	0
Total hardware static power p_c	30 dBm
The reciprocal of the power amplifier efficiency ψ_n	10
Time slot duration T	10 ms
Transmission latency T_l	50 time slots
Path-loss L	$-120.9 - 37.6 log10(dist)$
Shadowing attenuation ϕ	$\mathcal{N} \sim (0, 8^2)$
Batch size D	128
Discount factor η	0.3
Local-global experience replay buffer K	4000
Initial value of the noise variance ζ_{ini}	1.0
Minimum value of the noise variance ζ_{min}	0.05
Decay rate of the noise variance λ	0.9999
Learning rate of each critic DNN $\alpha^{(c)}$	0.001
Learning rate of each actor DNN $\alpha^{(a)}$	0.0001
Soft update rate of each target critic DNN $\tau^{(c)}$	0.001
Soft update rate of each target actor DNN $\tau^{(a)}$	0.001
(Target) Actor DNN structure	[s, 128, 256, 128, 1, 1]
(Target) Critic DNN structure	[s, 128, 128], [a, 128], [256, 1]
Activate functions of each (target) actor DNN	[Linear, relu, relu, relu, sigmoid, linear]
Activate functions of each (target) critic DNN	[Linear, relu, relu], [Linear, relu], [relu, linear]

over a duration of 2000 time slots. Meanwhile, the following results are the average of 10 trials with different seed settings.

4.5.1   Training Performance with Different Hyperparameters

In this part, we analyze the impact of various hyperparameters on the performance of the proposed algorithms. Our goal is to identify and select the most suitable parameters that optimize the algorithm's performance.

We first evaluate the learning rate of DNNs in achieving SE performance using the proposed spectrum-efficient MAIAC algorithm. From Fig. 4.5, the proposed MAIAC algorithms enhance the achieved SE reward across all the learning rate settings. However, when the learning rate of the DNNs is set to a much smaller value (e.g., 0.00001 and 0.0001, respectively), the proposed MAIAC algorithms take a much longer time to reach the optimal policy. Conversely, if the learning rate of the DNNs is set to a much higher value (e.g., 0.0005 and 0.005, respectively), the proposed MAIAC algorithms may be trapped in a suboptimal policy. To achieve a balance between the achieved reward and the learning speed, we set the learning rates of the actor DNNs and critic DNNs for both algorithms to $\alpha^{(a)} = 0.0001$ and $\alpha^{(c)} = 0.001$, respectively.

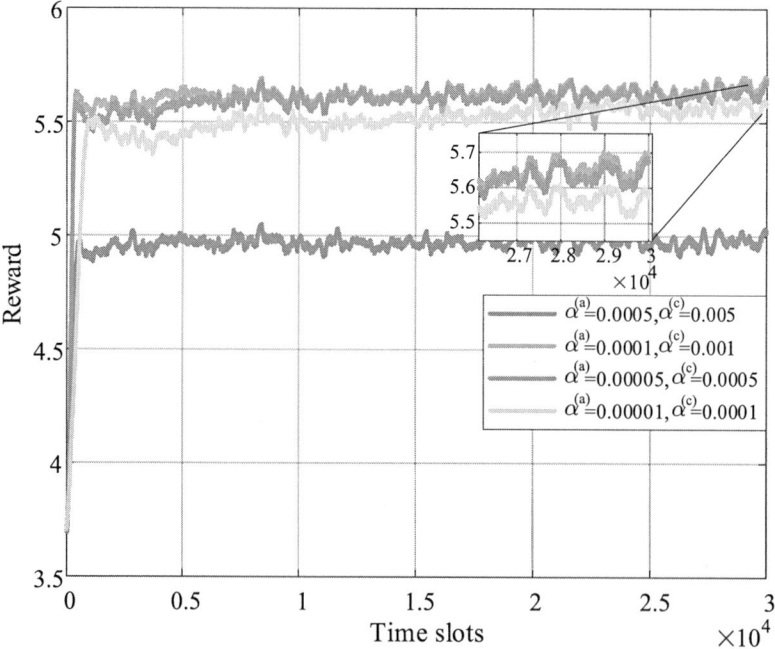

Fig. 4.5 Average global SE reward with different learning rates

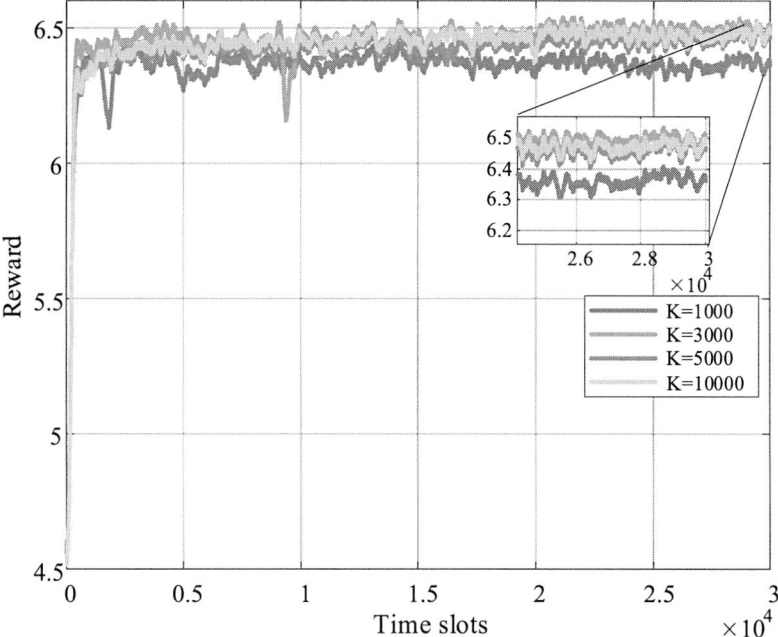

Fig. 4.6 Average global SE reward with different local-global experience buffer sizes

We then consider the effects of local-global replay buffer size on the performance. As demonstrated in Fig. 4.6, both too small (e.g., 1000) or too large buffer size (e.g., 10,000) can degrade the SE performance. Our results indicate that the optimal buffer size for spectrum-efficient MAIAC algorithm is 3000.

Finally, we look into the impact of the discount factor. Figure 4.7 reveals that a too small discount factor (e.g., 0.1) may limit the agents' view of the future and degrade the long-term reward, while a relatively higher discount factor in the range of $0.2 \sim 0.4$ enables the agents to balance short-term and long-term rewards and achieves a higher reward. Our results suggest that the optimal discount factor for spectrum-efficient MAIAC algorithm is 0.4.

4.5.2   Performance Comparisons in Quasi-Static Scenario

In this part, we aim to analyze the performance of SE in the quasi-static scenario. In particular, we implement the FP algorithm [8], the WMMSE algorithm [7], the *random power allocation* (RPA) algorithm, and the *maximum power allocation* (MPA) algorithm as benchmark algorithms for the SE performance evaluation.

Figures 4.8 and 4.9 provide the average SE performance. During the training phase, it is evident that the proposed spectrum-efficient MAIAC algorithm displays

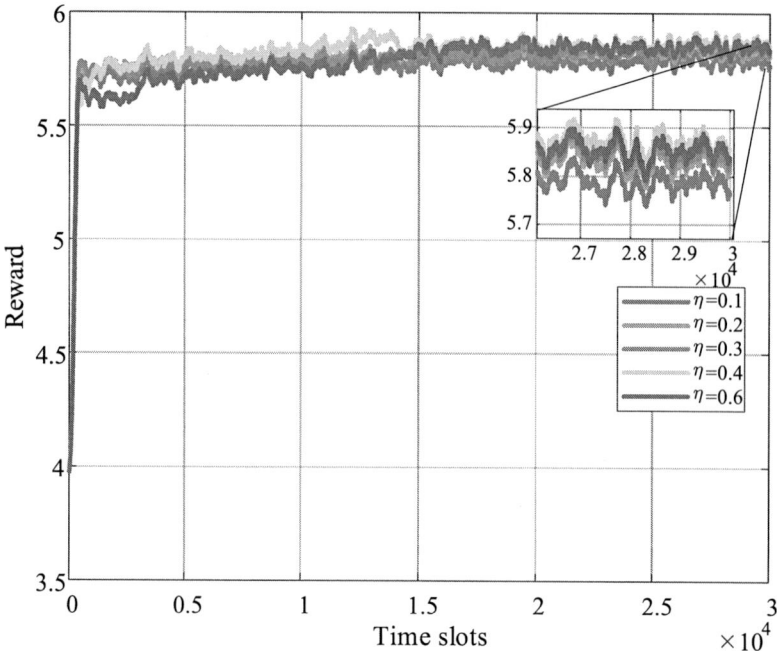

Fig. 4.7 Average global SE reward with different discount factors

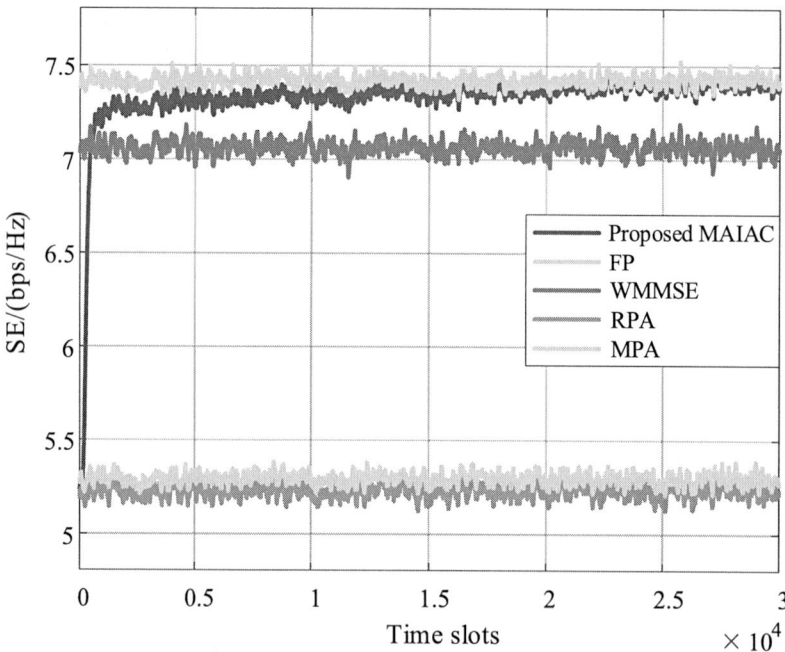

Fig. 4.8 SE performance comparisons in the training phase

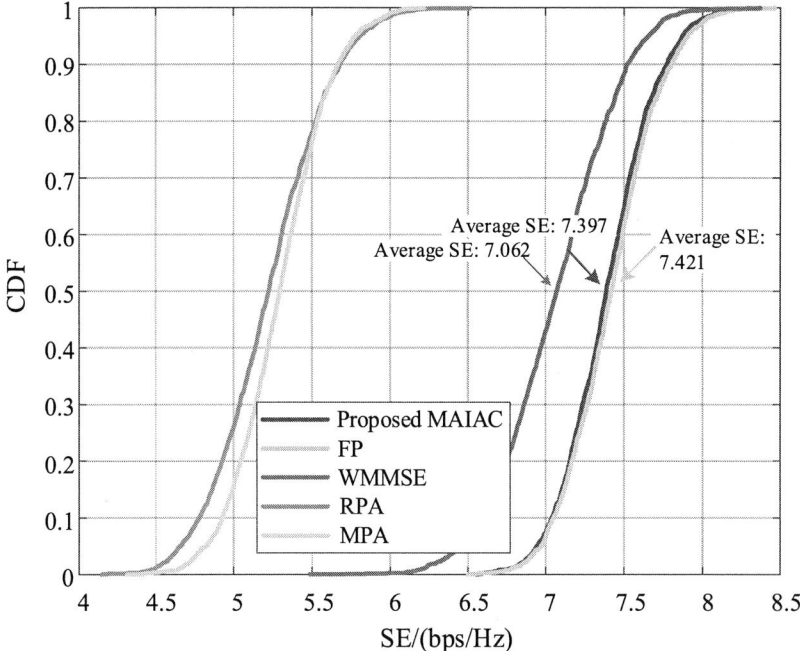

Fig. 4.9 SE performance comparisons in the testing phase

rapid convergence, and surpasses the WMMSE, the RPA and the MPA algorithms after approximately 1000 time slots. Moreover, the proposed spectrum-efficient MAIAC algorithm reaches the performance of the FP algorithm after around 20,000 time slots. During the testing phase, our results reveal that the proposed spectrum-efficient MAIAC algorithm achieves a performance of approximately 99.677% of the FP algorithm, while outperforming other benchmark algorithms even though the parameters of the actor DNNs are fixed, which illustrates that the trained actor DNNs is capable of handling unseen states.

Table 4.2 shows the training parameters of the DNNs and time complexity of the algorithms for SE optimization problem. Considering both (target) actor DNNs and (target) critic DNNs only take local information as input, which reduces the complexity of the input data. As a result, a relatively smaller number of training parameters is required to effectively train these DNNs. Besides, Table 4.2 shows the training time for each DNN is less than 2 ms, and for each target DNN is less than 3 ms. During the testing phase, each actor DNN takes around 0.793 ms to determine a transmit power, which is considerably faster than the time taken by both the FP algorithm (21.651 ms) and the WMMSE algorithm (42.172 ms) on average. These tables provide clear evidence of the superiority of the proposed MAIAC algorithms in terms of time complexity performance.

Table 4.2 Training parameters and time complexity comparisons

	Parameters	Training time (ms)	Testing time (ms)
Actor DNN	67,073	1.030	0.793
Target actor DNN	67,073	2.444	–
Critic DNN	83,841	1.829	–
Target critic DNN	83,841	2.591	–
FP [8]	–	–	21.651
WMMSE [7]	–	–	42.172

4.5.3 Performance Comparisons with Different Decorrelation Length

In this part, we investigate how the decorrelation length d_{corr} affects the overall performance of the algorithms in dynamic scenarios. For ease of notation, we define ξ_{PF} as the ratio of the proposed spectrum-efficient MAIAC algorithm's performance to that of the FP algorithm, ξ_{PW} as the ratio of the proposed spectrum-efficient MAIAC algorithm's performance to that of the WMMSE algorithm.

To begin with, we set v_n to 1 m/s for all UEs, and the large-scale shadowing attenuation becomes time-varying. Table 4.3 shows a comparison of the SE performance in the training phase. As $d_{corr} \in [10, 30]$ m, the proposed spectrum-efficient MAIAC algorithm takes around $22{,}000 \sim 24{,}000$ time slots for convergence, and can achieve a performance ratio of around $87 \sim 90\%$ compared to the FP algorithm and around $96 \sim 98\%$ compared to the WMMSE algorithm. As $d_{corr} \in [40, 50]$ m, the proposed spectrum-efficient MAIAC algorithm converges in less than 20,000 time slots and achieves a higher SE performance ratio, indicating that the proposed spectrum-efficient MAIAC algorithm is superior in dynamic scenarios with larger d_{corr}.

The performance ratio in the testing phase is depicted in Fig. 4.10, where both performance ratio curves increase as the decorrelation length increases. In particular, the proposed spectrum-efficient MAIAC algorithm achieves a performance ratio of approximately 93% compared to the FP algorithm and approximately 100% compared to the WMMSE algorithm with larger d_{corr}.

Table 4.3 SE performance with different decorrelation lengths in the training phase

d_{corr} (m)	Convergence overhead (time slots)	ξ_{PF}	ξ_{PW}
10	24,000	87.983%	96.210%
20	22,000	90.939%	98.943%
30	22,000	89.423%	98.529%
40	20,000	91.335%	97.988%
50	16,000	91.966%	99.291%

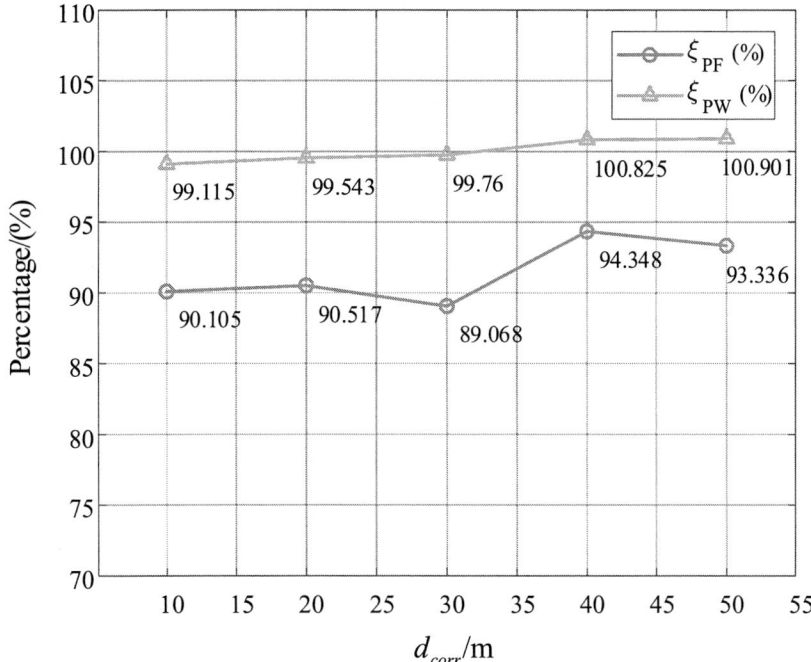

Fig. 4.10 SE performance ratio with different d_{corr} in the testing phase

4.5.4 Performance Comparisons with Different UE's Velocities

In this part, we set d_{corr} to 20 m and evaluate how the UE's velocity v_n affects the performance.

Table 4.4 provides a comparison of SE performance in the training phase. As v_n increases from 2 to 10 m/s, the convergence overhead gradually increases. This is because the wireless channel becomes more dynamic as v_n increases, such that more time slots are needed for the proposed spectrum-efficient MAIAC algorithm to learn the environment pattern. Additionally, the proposed spectrum-efficient MAIAC algorithm achieves a performance ratio of around $90 \sim 91\%$ compared to the FP algorithm and around $96 \sim 98\%$ compared to the WMMSE algorithm when v_n ranges from 2 to 6 m/s. However, as v_n exceeds 6 m/s, the performance ratios decrease slightly by approximately $2 \sim 4\%$. This suggests that the proposed spectrum-efficient MAIAC algorithm is more adaptable to the low-speed scenarios than high-speed scenarios.

Figure 4.11 illustrates the SE performance ratio in the testing phase, it can be observed that both performance ratio curves gradually decrease as v_n increases. In particular, the proposed spectrum-efficient MAIAC algorithm can still achieve a performance ratio of approximately 90.842% compared to the FP algorithm and

Table 4.4 SE performance with different UE's velocities in the training phase

v_n (m/s)	Convergence overhead (time slots)	ξ_{PF}	ξ_{PW}
2	23,000	91.581%	96.883%
4	24,000	90.968%	98.454%
6	27,000	90.143%	98.226%
8	29,000	89.791%	94.705%
10	28,000	90.292%	93.725%

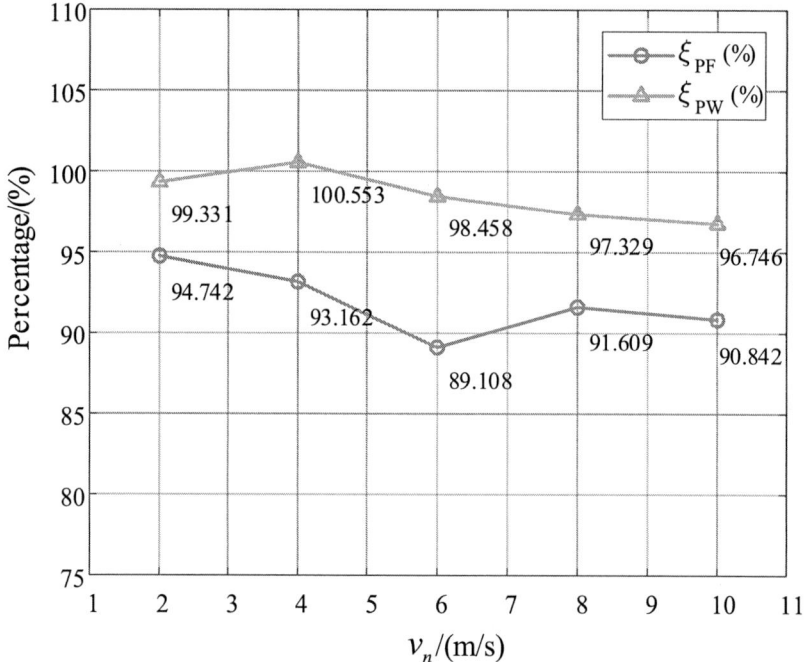

Fig. 4.11   SE performance ratio with different v_n in the testing phase

approximately 96.746% compared to the WMMSE algorithm when v_n is set to 10 m/s, which is higher than that in the training phase.

4.6   Conclusions

In this chapter, we proposed a lightweight cloud-edge collaboration framework by enabling lightweight collaborations among the cloud and different edge BSs in a typical wireless edge network. With lightweight collaborations, the edge BSs do not need to exchange local instantaneous information with each other and the cloud collects only data rate from edge BSs to compute global historical SE, which is then fed back to edge BSs. Within the framework, we introduced spectrum-efficient MAIAC power control algorithm, thus each BS can gradually learn to optimize local power control policy to improve global SE in a distributed manner. Simulation results demonstrate that the proposed spectrum-efficient MAIAC algorithm can achieve over 99% SE performance of conventional algorithms in quasi-static scenarios, and approximately 89 $\sim$ 100% in dynamic scenarios with significantly reduced time complexity.

References

1. Zhang, L., Liang, Y.-C., Niyato, D.: 6G visions: mobile ultrabroadband, super Internet-of-Things, and artificial intelligence. China Commun. **16**(8), 1–14 (2019)
2. Alsabah, M., Naser, M.A., Mahmmod, B.M., et al.: 6G wireless communications networks: a comprehensive survey. IEEE Access **9**, 148191–148243 (2021)
3. Xu, Y., Gui, G., Gacanin, H., Adachi, F.: A survey on resource allocation for 5G heterogeneous networks: current research, future trends, and challenges. IEEE Commun. Surv. Tutor. **23**(2), 668–695 (2021)
4. Yang, C., Li, J., Guizani, M., Anpalagan, A., Elkashlan, M.: Advanced spectrum sharing in 5G cognitive heterogeneous networks. IEEE Wirel. Commun. **23**(2), 94–101 (2016)
5. He, A., Wang, L., Elkashlan, M., Chen, Y., Wong, K.-K.: Spectrum and energy efficiency in massive MIMO enabled HetNets: a stochastic geometry approach. IEEE Commun. Lett. **19**(12), 2294–2297 (2015)
6. Gao, H., Wang, M., Lv, T.: Energy efficiency and spectrum efficiency tradeoff in the D2D-enabled HetNet. IEEE Trans. Veh. Technol. **66**(11), 10583–10587 (2017)
7. Shi, Q., Razaviyayn, M., Luo, Z.-Q., He, C.: An iteratively weighted MMSE approach to distributed sum-utility maximization for a MIMO interfering broadcast channel. IEEE Trans. Signal Process. **59**(9), 4331–4340 (2011)
8. Shen, K., Yu, W.: Fractional programming for communication systems-Part I: Power control and beamforming. IEEE Trans. Signal Process. **66**(10), 2616–2630 (2018)
9. Lee, W., Kim, M., Cho, D.: Deep power control: transmit power control scheme based on convolutional neural network. IEEE Commun. Lett. **22**(6), 1276–1279 (2018)
10. Zhou, F., Zhang, X., Hu, R.Q., Papathanassiou, A., Meng, W.: Resource allocation based on deep neural networks for cognitive radio networks. In: Proceedings of the IEEE/CIC International Conference on Communications in China (ICCC), pp. 40–45 (2018)
11. Mnih, V., et al.: Human-level control through deep reinforcement learning. Nature **518**, 529–533 (2015)
12. Gu, S., Holly, E., Lillicrap, T., Levine, S.: Deep reinforcement learning for robotic manipulation with asynchronous off-policy updates. In: Proceedings of the IEEE International Conference on Robotics and Automation (ICRA), pp. 3389–3396 (2017)
13. Niroui, F., Zhang, K., Kashino, Z., et al.: Deep reinforcement learning robot for search and rescue applications: exploration in unknown cluttered environments. IEEE Robot. Autom. Lett. **4**(2), 610–617 (2019)
14. Chu, T., Wang, J., Codeca, L., Li, Z.: Multi-agent deep reinforcement learning for large-scale traffic signal control. IEEE Trans. Intell. Transp. Syst. **21**(3), 1086–1095 (2020)
15. Xiao, Y., Xiao, L., Wan, K., Yang, H., Zhang, Y., Wu, Y., Zhang, Y.: Reinforcement learning based energy-efficient collaborative inference for mobile edge computing. IEEE Trans. Commun. **71**(2), 864–876 (2023)
16. Nasir, Y.S., Guo, D.: Deep actor-critic learning for distributed power control in wireless mobile networks. In: Proceedings of the Asilomar Conference on Signals, Systems, and Computers (ACSSC), pp. 398–402 (2020)

17. Giannopoulos, A., Spantideas, S., Kapsalis, N., Karkazis, P., Trakadas, P.: Deep reinforcement learning for energy-efficient multi-channel transmissions in 5G cognitive HetNets: centralized, decentralized and transfer learning based solutions. IEEE Access **9**, 129358–129374 (2021)
18. Iqbal, A., Tham, M.-L., Chang, Y.C.: Energy- and spectral- efficient optimization in cloud RAN based on dueling double deep q-network. In: Proceedings IEEE International Conference on Automatic Control and Intelligent Systems (I2CACIS), pp. 311–316 (2021)
19. Liu, Z., Chen, X., Chen, Y., Li, Z.: Deep reinforcement learning based dynamic resource allocation in 5G ultra-dense networks. In Proceedings of the IEEE International Conference on Smart Internet Things (SmartIoT), pp. 168–174 (2019)
20. Sun, H., Chen, X., Shi, Q., Hong, M., Fu, X., Sidiropoulos, N.D.: Learning to optimize: training deep neural networks for interference management. IEEE Trans. Signal Process. **66**(20), 5438–5453 (2018)
21. Zhang, L., Liang, Y.-C.: Deep reinforcement learning for multi-agent power control in heterogeneous networks. IEEE Trans. Wirel. Commun. **20**(4), 2551–2564 (2021)
22. Peng, J., Zheng, J., Zhang, L., Xiao, M.: Deep reinforcement learning for energy-efficient power control in heterogeneous networks. In: Proceedings of the IEEE International Conference on Communications (ICC), pp. 141–146 (2022)
23. Zhang, L., Peng, J., Zheng, J., Xiao, M.: Intelligent cloud-edge collaborations assisted energy-efficient power control in heterogeneous networks. IEEE Trans. Wirel. Commun. **22**(11), 7743–7755 (2023)
24. Chaipanya, P., Uthansakul, P., Uthansakul, M.: Performance enhancement employing vertical beamforming for ffr technique. J. World Acad. Sci. Eng. Tech. **7**, 478–481 (2013)
25. Kim, T., Love, D.J., Clerckx, B.: Does frequent low resolution feedback outperform infrequent high resolution feedback for multiple antenna beamforming systems? IEEE Trans. Signal Process. **59**(4), 1654–1669 (2011)
26. Luo, Z.-Q., Zhang, S.: Dynamic spectrum management: complexity and duality. IEEE J. Sel. Topics Signal Process. **2**(1), 57–73 (2008)
27. Lillicrap, T.P., Hunt, J.J., Pritzel, A., Heess, N., Erez, T., Tassa, Y., Silver, D., Wierstra, D.: Continuous control with deep reinforcement learning. In: Proceedings of the ICML, pp. 1–14 (2016)

Chapter 5
Conclusions

Abstract This chapter concludes the Brief. Since mobile data traffic has kept increasing in recent years, spectrum sharing has emerged as a promising way to alleviate the contradiction between the limited spectrum resource and the increasing mobile data traffic in wireless edge networks. In this Brief, we investigate the AI-enabled spectrum sharing in wireless edge networks and focus on different spectrum sharing scenarios. Specifically, we leverage reinforcement learning to enhance communication performance through several proposed methods, i.e., DQN-based MCS selection method for opportunistic scenarios, DDPG-based power control method for centralized scenarios and DDPG-based power control method for distributed scenarios. In particular, our proposed methods could be further studied in more practical situations by considering the convergence time in various scenarios and the synchronization designs, and has the potential to be extended into space-air-ground integrated networks in the future.

Keywords Opportunistic spectrum sharing · Centralized spectrum sharing · Distributed spectrum sharing

Recent years have witnessed the explosive growth of mobile data traffic in wireless edge networks, and the mobile data traffic therein is predicted to further increase in the next decades. The contradiction between the limited spectrum resource and the increasing mobile data traffic will be more and more apparent. Spectrum sharing has been proven as a promising way to alleviate the contradiction in wireless edge networks.

The successful applications of AI in computer and control areas motivate the exploration of the AI-enabled spectrum sharing, which has attracted extensive attention from global scholars in recent years. In this Brief, we investigate the AI-enabled spectrum sharing in wireless edge networks and focus on opportunistic, centralized, and distributed spectrum sharing scenarios. In the opportunistic spectrum sharing scenario, primary and secondary links share the same spectrum resource. We proposed a DQN-based MCS selection method to protect primary links from the interference caused by the secondary links and enhance the throughput performance of primary transmissions. In the centralized spectrum sharing scenario, multiple BSs share the same spectrum resource to serve their associated users. We proposed a centralized DDPG-based power control method, such that each BS can automatically configure its transmit power to enhance the network spectrum efficiency. In the distributed spectrum sharing scenario, multiple BSs share the same spectrum resource to serve their associated users. We proposed a distributed DDPG-based power control method, such that each BS can adapt its transmit power intelligently to boost the network spectrum efficiency.

This Brief mainly leverages the reinforcement learning for AI-enabled spectrum sharing. There are two main drawbacks to applying reinforcement learning in wireless edge networks. The first one is the convergence time. Different scenarios and algorithm designs may lead to various convergence times. If the convergence time is long, the costs may be unacceptable in terms of the performance degradation. Then, the digital twins technique can be a promising alternative by creating a virtual wireless edge network and learning an efficient spectrum sharing policy offline. The other drawback is the synchronization among multiple agents. The existing multi-agent reinforcement learning based spectrum sharing methods typically assume perfect synchronizations among multiple agents. In fact, it is challenging to guarantee perfect synchronizations in practical situations. Then, robust multi-agent reinforcement learning based spectrum sharing methods are needed by taking the synchronization issue into considerations.

Besides, the wireless edge network is evolving toward the integrated air-space-ground network, in which the spectrum sharing is still a key way to deal with the spectrum shortage issue. Nevertheless, the spectrum sharing in an integrated air-space-ground network is more complicated than conventional wireless edge network. Thus, there are many open issues to be addressed in the AI-enabled spectrum sharing of the integrated air-space-ground network.